MW00850126

THE NEW ORDER

How AI Rewrites the Narrative of Science

CHRIS EDWARDS

 Prometheus Books

Essex, Connecticut

 Prometheus Books

An imprint of The Globe Pequot Publishing Group, Inc.
64 South Main Street
Essex, CT 06426
www.globepequot.com

Distributed by NATIONAL BOOK NETWORK

Copyright © 2025 by Chris Edwards

All rights reserved. No part of this book may be reproduced in any form or by any electronic or mechanical means, including information storage and retrieval systems, without written permission from the publisher, except by a reviewer who may quote passages in a review.

British Library Cataloguing in Publication Information available

Library of Congress Cataloging-in-Publication Data is available

ISBN 9781493089116 (cloth) | ISBN 9781493089123 (ebook)

♾️™ The paper used in this publication meets the minimum requirements of American National Standard for Information Sciences—Permanence of Paper for Printed Library Materials, ANSI/ NISO Z39.48-1992

CONTENTS

Introduction 1

**Part I: The Old Order: Section Introduction: Philosophy and
Artificial Intelligence** 13

A   The Pre-Socratics 17

B   Socrates, Plato, and Aristotle 21

C   The Hellenistic Era .27

D   Rome and Lucretius .31

E   The World Historical Narrative 35

F   The Theologians 39

G   Gunpowder, the Compass, the Clock, and Block Printing 47

H   Numbers and the Black Plague 53

I   Descartes, Musical Theory, and the X-Y Graph 55

J   Galileo and Bacon 59

K   Isaac Newton 65

L   Binary Code and the Law of Large Numbers 73

M   The Great Era of Mathematics 77

N   Revolutions and the Carnot Cycles 81

O   Lavoisier and the Law of Conservation 91

P   Rudolf Clausius, Energy, and Entropy 95

Q   James Clerk Maxwell 101

R   The Curies and a Demon 105

S   Maxwell and Electromagnetism 109

T   The Pre–Quantum Mathematics Crisis 113

U   Einstein 121

V   Wittgenstein and an Incomplete Circle 131

W   Planck, Bohr, the Curies, Heisenberg, and Uranium 139

X   Uncertainty and Exclusion 145

Y   Convergence: Relativity and Thermodynamics 153

Z   The Third Law of Thermodynamics, String Theory,
    and Model-Dependent Realism 159

Part II: The New Order: Section Introduction: Bend Sinister . . **167**
Unifying the Scientific Narrative
    under the Laws of Thermodynamics 199

Appendix 1: "There Is Only Entropy: Unifying the Narrative of
    Science" 201

Appendix 2: "Probability and Waves: A Synthesis of
    Three New Books" 209

References 217

About the Author 221

INTRODUCTION

WHAT IF THE GREAT DISCOVERIES OF SCIENCE CAME IN THE WRONG ORDER? The assumptions in the question almost overwhelm the overall conceit presented by the question itself. It is assumed that there is such a thing as a "great" discovery. Also, by asking whether those great discoveries came in the "wrong" order, there follows an implication that there must be a "right" order for them to have occurred in, and what does that mean?

Since this book is an attempt to answer this one-sentence question, it is crucial that the question itself be fully explained and understood. This requires a parsing out of the assumptions, beginning with the concept of a "great discovery."

The scientific method cannot be understood as discrete from historical and social developments. Science sometimes proceeds on the idea that the scientific method was always there waiting to be found. In a world historical context, the Indians, Chinese, and Muslims discovered the scientific method but, without a printing press, could not encode the concept into their civilizations. Only in the West, for a variety of reasons, could science both be discovered and, because of the printing press, be made to "stick" in the culture.

The problem with this framework of understanding is that it divorces the natural philosopher/scientist from the science. Scientific theorists cannot be separated from the historical and linguistic forces that shaped the way in which they pose questions and theorize answers. When novel ideas develop into language, the words for those ideas create a powerful shaping effect on the mind.

In his masterful work *The Invention of Science: A New History of the Scientific Revolution* (2016), David Wootton wrote of how a language evolved in Western Europe that was then adapted by science (what was then called "natural philosophy"). The word "discovery" cannot be found in the Western lexicon until after Columbus sailed west across the Atlantic in 1492. The new lands and peoples he found there created a new conceit: that new things could be *found* out there in the world. In 1620, when Francis Bacon sought to codify the scientific method into education, he chose the image of a ship sailing through the Pillars of Hercules for the cover of his *Novum Organum* (new organ, or new method).

The concept of discovery and the accompanying sailing analogy then synthesized with Western philosophy's oldest questions about the nature of matter: Aristotle's concepts of motion and Plato's ephemeral conceits about mathematics. Combined with Arabic numerals and the pressing questions of physics that cannon technology created, this environment eventually produced an Isaac Newton, among other notables.

In order to proceed in understanding the ideas in this book, the reader will need to accept the concept that individuals are born into societies pre-shaped by previous philosophies, ideas, and theories and that science is not discrete from this pre-shaping. By analogy, your average American high school student might not know anything about the Protestant Reformation or know the name of Martin Luther. However, if one student is caught cheating on a math test and the teacher fails the entire class as a result, there would very likely be an outcry from the bystander students (and their parents).

The reasons for this is that our Western educational institutions, like our criminal justice system, have been shaped over five centuries, during which time the concept of individual judgment became a central prerogative of Western philosophy, education, and jurisprudence. That is not to say that every law in Western civilization can be traced back to a Protestant ideal. Rather, the Protestant conceit of individual judgment was adapted into Enlightenment philosophy and then into governments in the same way that "discovery" was adapted from a maritime language into that of natural philosophy.

To use another example, Sigmund Freud seems to have inalterably changed the way that people in Western civilization think about childhood trauma. If a hypothetical adult dislikes or is afraid of dogs, then she might explain her revulsion as being the product of having been attacked in childhood by a dog. This is not to say that everyone who dislikes dogs would say such a thing or that everyone who is attacked by a dog as a child will grow into adulthood with that fear or dislike of dogs. It is simply to say that in Western civilization, adults feel comfortable connecting their behaviors to childhood influences and/or traumas. Neither generations of skeptical assaults on Freud's methods nor modern theories about behavior and genetics have made it broadly accepted for an adult to say, "I dislike dogs because I am genetically predisposed to do so."

To broaden the concept a bit, we might understand William Shakespeare's beloved *Romeo and Juliet* (1595) not only as being a creation from the mind of a genius but also as having been pre-shaped by a Western culture where the literature had traditionally favored individuality and rebelliousness. The Greeks wrote of individual heroes like Achilles or the doomed Antigone. Rebellion against authority evolved into a virtue. Shakespeare's audience understood the conceit of star-crossed lovers who acted in defiance of their parents.

However, no Chinese authors could have produced a *Romeo and Juliet* because China was shaped by Confucian philosophy and filial piety. For young lovers to defy their parents or for a young person to assert that he or she should be the one to choose his or her life mate would have been seen as deeply sinful in 16th-century China, as inconceivable to them as a gay and vegetarian cowboy would have been to the original audience of *Bonanza*.

With this rationale in place, we can surmise the historical shaping effects on science in Western civilization. Pre-Socratic philosophers asked, "What is the nature of matter?," not "Why does everything disintegrate over time?" Then Aristotle theorized about the physics of motion, the numbers 0 through 9 made it possible to calculate incremental change, and the cannon provided a direct example of the quadratic equation. *And then*, in 1666, when a young Isaac Newton (probably apocryphally) looked up at the apple tree, he asked why the apple always fell, not why it always rotted.

Now we come to the second assumption from our introductory question. What does it mean for a discovery to come in the wrong order?

3

Scientists across disciplines can be assured, with absolute certainty, of one thing: When scientists stare out at the supernova, squint into a microscope at a zygote, or study ancient metals buried in the earth, it becomes apparent, no matter what the scientific object is, that everything decays.

And that decay creates heat. And that is called entropy, radiation, or the second law of thermodynamics.

Isaac Newton did not know about this central theorem. This is probably because the thinking of Western civilization's ancient philosophers was constrained by their minute life spans. A human life span allows us to see that both a piece of iron and an apple, if dropped, will go down. We cannot see, over the span of a human life, that both the iron and the apple rot.

To understand why this is an issue for creating a scientific narrative, we need to understand the theory of collocation. The *Dictionary of Scientific Theories* (2001) states this:

> Collocation (1951). *Linguistics* Idea proposed by the British linguist John Rupert Firth (1890–1960).
>
> The distributional tendency of certain words to appear close together—"collocate"—in texts: 'dark' and 'night,' 'time' and 'save,' 'spend' and 'waste,' and so on. Firth claims that the range of a term's collocates is part of its meaning.
>
> —J. R. Firth, "Modes of Meaning," *Papers in Linguistics 1934–1951* (London, 1957).

Ideas turn into words, and then words synthesize into conceits and concepts. Firth's genius was in finding a way to demonstrably measure this phenomenon by looking at immobile words on the paper. When words are distributed close together, this indicates a collocation and therefore an extension of meaning.

Richard Dawkins had a similar idea with his biological concept of an "extended phenotype." A dam, to use his best example, can be considered an extension of the beaver's physical presence. Also, George Lakoff and Mark Johnson's classic work *Metaphors We Live By* (1980) contained a section where the authors argued that "time" and "money" are thought of in the same way: both are "wasted," "saved," and "spent." It is hard for people in a capitalist economy not to think of a day as being 24 monetary units.

Scientifically speaking, once an idea becomes a word and once those words collocate with other words and then get reproduced in textbooks, scholarly journals, and presentations, over time, the words and ideas meld into one another. For example, whenever I point out to people that a clock is just a ruler bent into a circle, they almost always respond with some degree of shock. Here is one of the most fundamental connections in Western intellectual life: the concepts of space and time, interlinked by two of the most common and enduring objects in everyday life, and almost no one ever sees it without direction.

Pointing out the similarity between the clock and the ruler is not trivial. Your experience of distance changes depending on your speed, and so does your experience of time. A ruler measures distance, and a clock, if you think about it, measures movement. Or, rather, it keeps a steady movement in a universe where movement is chaotic. If time measures movement, then if there is no movement, there is no time. (Hence, the equation $t = 0$ indicates the singularity.)

To use a nonscientific example, consider the phrase "trench coat." The phrase probably conjures up an image of a man in a long Dick Tracy–style coat. Generally, someone wearing a trench coat is depicted as a businessman, a gangster, or an enforcer of some kind. The trench coat, however, originated in World War I. Soldiers wore them as protection against the rats and the elements. Soldiers wore them in the trenches. When the trench coat first became a part of society, the meaning was likely immediately apparent. Over time, the word "trench" melted into the fabric of the word and simply became a fashion term disconnected from its original meaning.

"Wrong order" therefore means that scientific discoveries came in such a way that ideas and the nomenclature created by those ideas impeded the understanding of later ideas and the understanding of those ideas. The result is a confusion of concepts, often understood only by specialists, that at certain levels of thought end up in contradiction. To use two examples, the law of conservation in chemistry indicates that matter and energy cannot be either created or destroyed in an experiment. In quantum mechanics, however, there is a statistical guarantee that if you perform enough experiments, there will be a heat loss. The universe overall will not gain or lose anything, but the experiment will.

Or, in mathematics, there exists the basic commutative property, where $A \times B = B \times A$. In quantum mechanics, however, interactions are not commutative, so this mathematical property must be discarded in favor of matrix algebra. Given what we now know about interactions and heat loss, the commutative "property" might be better described as a fallacy.

The reason for this is likely because the word "energy" became popular among scientists in the early 19th century. "Energy" has a positive connotation as a force to be used for growth, order, and power. Entropy, as a concept, developed only with the development of quantum mechanics, and although heat loss was discovered by the Sadi Carnot in the early 19th century, the concept of entropy was not really established until the creation of quantum mechanics in the early 20th century. By then, the E in $E = mc^2$ was already established as energy. But what the E really stands for is "entropy."

The confusion comes from the definitions of "entropy" and "energy." These words have been used as synonyms, altering only when specialists look at the subject. Unifying the scientific narrative depends on the clarification of these definitions:

A. Entropy: This is the decay of a particle, the result of internal movement that creates heat. Heat loss comes in a variety of forms, fundamentally with the release of beta particles, but the result is always radiation. Radiation can be described in a variety of forms as alpha, beta, or gamma radiation and is released at different levels of intensity. At certain levels, the radiation takes on different forms, such as gluons or quarks, but those details cannot affect the fact that overall level of entropy in the universe always increases.

B. Energy: Energy is one object's temporary capture of another object's entropy for the purpose of creating temporary order. Energy is a secondary force to entropy. Entropy can be described as energy only when that force is used to create temporary order.

Tautologies haunt all definitions, but these two above should be considered as laws. *All* forms of energy are the result of some object's decay. A nuclear scientist studying the sun, for example, will describe the process of thermonuclear fusion as the radiation of hydrogen, and the decay creates hydrogen isotopes and heat. The sun's entropy radiates out into the solar system. A horticulturist will then describe the sunlight as energy when it is absorbed by plants through photosynthesis. The "energy" here can come only from the sun's decay. The nuclear scientist and the horticulturist study the same force with different purposes and different nomenclature, and this causes confusion for anyone trying to engage in cross-curricular study.

When someone charges a cell phone, she is borrowing the decay of coal or oil, which is then funneled into her home or car in the form of electricity. The electricity is "entropy-become-energy" and is used to force herd decaying particles in the battery back into a sense of order. Unplugging the phone releases the energy holding back the particles in the battery, and then those particles begin the natural process of decay. That decay releases heat (which can be felt), which the apps on the phone treat as energy until the process of entropy is complete and the phone "dies" just as sure as the sun will do one day, just as surely as everything will.

So this is partially what is meant by the great scientific discoveries coming "in the wrong order." If entropy had been described before gravity and if 19th-century chemists had understood entropy as the primary force and energy as a secondary and temporary force, then everything in science would coalesce and make more sense.

Because any book about science is, in some ways, a book about science education, we have come to the point of this book. The objective here is to break down the traditional scientific narrative, with the herky-jerky progress of science through discoveries and theories that came in an arbitrary order, and to put those discoveries into a narrative format that makes sense for readers, scientists, educators, and students. In order to achieve this objective, readers should be aware of the parameters of the book, described next.

There are no new equations in this book, and the intent is not to create a "theory of everything." The purpose of the book is to show that all of the major breakthroughs in science and all of the important theoretical constructs can be explained more efficiently using the definitions of entropy and energy that I have created here. It does no violence, for example, to $E = mc^2$ to describe the E as standing for "entropy" rather than "energy." By defining the E as "entropy," however, the educational narrative of science across disciplines becomes more comprehensible.

Thomas Kuhn's classic *The Structure of Scientific Revolutions* (1962) created the concept of the scientific "paradigm." In Kuhn's context, the word "paradigm" meant the overarching scientific worldview. People in late medieval and early modern Europe believed Earth to be at the center of the universe (understood then as being bound by what the eyes could see in the night sky). Copernicus, Galileo, and the telescope put that geocentric paradigm in a state of crisis, and eventually the geocentric system was deconstructed. A heliocentric structure was constructed in its place.

Kuhn's thesis, always interesting, frequently comes in useful for theoretical physicists. The reason for this is because the geocentric model was not wrong. Geocentric believers could describe existing phenomena (the positions of the sun, Mercury, Mars, Venus, and Jupiter) and make clear predictions (that the sun would "rise" and "set" tomorrow). However, the geocentric theorists could not describe the motion of the moons of Jupiter that Galileo had seen through his telescope. The heliocentric theory could describe everything that the geocentric theory could in addition to the motion of those moons and much more as well.

Unfortunately, Kuhn's thesis has permeated into scientific and pseudoscientific culture. "Paradigm" is now used in discussions regarding new visions for a company and is popular largely because it allows anyone with a new idea to frame himself as a heroic Galileo battling against the guardians of an old and failed order. Too many books of science are now framed by Kuhn's conceits, with an author seeking to overturn existing theories and to engage in a Hegelian warfare of ideas against an old synthesis.

This is all annoying and unproductive. I like books and ideas and prefer to engage in discussions with other people who like books and ideas so that better books and ideas can be created through the process. To begin an intellectual endeavor with the intent of tearing down an existing order or to insinuate (as is too often the case) that old paradigms cease to exist only when the intellectuals who created them get old and die is just not a good way to start a conversation.

It is not necessary in this book for me to declare anyone to be *wrong* about anything. No paradigms will be torn down or built. I hope that no one, particularly any valuable old theorists, dies. Rather, the intent here is to show how the full brilliance of a great many scientific theories can be understood only when the narrative is united and simplified. My hope is to present the scientific narrative in such a way that it illuminates the full genius of many of the great figures of science.

A few words are in order about my own background that are relevant to the writing of this book. I write frequently for *Skeptic* magazine (among many other publications) on a variety of topics, including logic, theoretical physics, anthropology, psychology, history, education, and philosophy. My special background—and the primary force of my publications—is in world history, a field that I assert is not *a subject* but rather *all the subjects*. World history, a young field, requires cross-curricular study and the development of broad narrative connections.

In my next-to-last book, *Femocracy: How Educators Can Teach Democratic Ideals and Feminism*, the central thesis was that Western civilization's central story really is about the development of feminism and the emergence of feminine virtues as being encoded into Western society. The narrative of Western civilization, goes the argument, both makes more sense and provides greater guidance for students of history when feminism is at the center of the story. That is a world historical argument, not necessarily one that a traditional historian, interested in, say, the development of the Prussian infantry, could make.

To that end, I should note that a world historian is not a researcher but a synthesizer. My books are historical essays about simplifying narratives and creating interesting pathways through knowledge. This is the goal of education.

Something similar will be done in this book. This is not a work of science or mathematics in the sense that no new research will be composed. However, the simplifying of the scientific narrative around the second law of thermodynamics will require that the traditional narrative be broken down, rearranged, and then re-created. Just a few sections of this book, especially the section on Einstein, will be based on some of my previously published work. My hope is that readers will sympathize that this book deals with a number of complicated ideas, and, where possible, I sought peer review and publication as verification for my interpretations.

I am not a mathematician, but I am not in awe of mathematicians either. The greatest and most famous theoretical physicists create probable postdictions about the universe and then construct elaborate narratives around those mathematical probabilities. Lay audiences often get lost in the colored smoke that comes from all the conjuring, but I know the tricks that go on behind the curtain, and some of them lead the mathematicians astray. I do admire mathematics and am a fair self-taught practitioner of the art, but mathematicians have had their go at unification theories. For the inherent purpose of this book, other methods beyond mathematics are needed.

To that end, this book will be divided into two unequal parts. The first and longer part will be a history of science told in the traditional order. *This is not to say that part I will be a traditional scientific narrative*, as my intent is to narrate this story with a few twists. When theories and discoveries are presented in their original orders, in the place of chapters, an A–Z format will be used so that I can reference each scientific idea in part II. Part I will end with the discovery of Hawking radiation and the crucial understanding that even black holes, the entities with the highest concentrations of gravity, are subject to entropy. The original paper, published in *Skeptic*, is included in the appendix.

Part II will explain the centralized place of entropy in the scientific narrative. The second law of thermodynamics will be to part II what Hamlet was to *Hamlet*. The letters used in part I will be brought up again here so that each discovery and theory can be described in terms of entropy. These descriptions will not change any individual theory but will make the interconnectedness to the theories apparent.

Overall, although my goal is unification and simplification, this book might become complicated in places. I enjoy reading books and like giving credit to authors for either their original ideas or their explanations. I will include a few lengthy sections from the books of other authors, and I do that for a few reasons: to avoid plagiarism and patchwriting and to draw attention to the good work of other authors, to direct readers to a significant moment in the history of science by highlighting an original source, and to show sometimes that the original sources were written in an opaque form that led to a later misunderstanding.

I enjoy reading, and to have been able to draw on the work of so many talented thinkers was the central pleasure of writing this book. To make it easier for readers to continue on with the sources that I drew from, references are included at the end. The narrative presented in part I is well known, so I decided not to disrupt the flow of the writing by placing references after each letter. In part II, the narrative will be presented with each letter from part I referenced in parentheses when applicable and necessary.

My hope is always to be clear in the narrative; however, in this book, there will always be a single conceit at the center of the thesis, and it is one that artificial intelligence will probably begin with if and when it rewrites the history of science with a more direct connection with theory and mathematical symbology.

The central thesis is simple, unavoidable, and universal: the laws of thermodynamics reside at the core of science. If we reshuffle the order of science's great discoveries, then all of the important innovations can be understood as variations of thermodynamic properties, particularly the second law of thermodynamics. In part I, those discoveries will be presented in their chronological order and then explained in the context of thermodynamic properties; then in part II, they will be reshuffled into a new order.

PART I

THE OLD ORDER

Section Introduction: Philosophy and Artificial Intelligence

EXPERTS IN ARTIFICIAL INTELLIGENCE (AI) (AND I AM NOT ONE) TEND to focus on AI's potential uses in various fields. My expertise is primarily in two fields: (A) the history and philosophy of science (as an important subheading to world history) and (B) education, particularly in the power of cross-curricular study (where world history seems to be the perfect nexus). Below, I will succinctly detail AI's potential in both fields and its connections with my previous work.

A. THE HISTORY AND PHILOSOPHY OF SCIENCE

The details will be developed in the text of part I, but philosophers have long been obsessed with the notion of solipsism. All of us are limited to one mind. I cannot tell if other individuals perceive "red" or "cold" in the same way that I do. The notion of solipsism has been expressed variously as the "thing-in-itself" by Immanuel Kant or as "atomic facts" by Ludwig Wittgenstein. Both thinkers, however, inverted the solipsism by placing the question on objects that give off sensory data (anything we can smell, touch, see, or hear).

Gradually, scientists and philosophers have come to understand that humans evolved consciousness under evolutionary principles, that is, that our mental models help us to survive and mate and thus are only tangentially connected to atomic reality. AI will be able to directly understand objective sensory data, and all sensory data can be defined in a more direct manner as a species of radiation. That brings us to AI's function in education.

B. Education

By "education," I am not referring to AI's potential for instructing students. I am referring to AI's ability to learn in a different way and to think for itself in a way unconstrained by human history. To begin, the human brain is preadapted but poorly suited for higher-order thinking. AI will not need metaphors or analogies. To learn math, humans must spend years painfully trying to download mathematical formulas onto neurons that actively seek to reject the process in favor of more pleasurable stimulations. Machines must "learn" formulas only once, and the understanding is instant and perfect.

Humans face all sorts of practical problems when it comes to learning. Most humans consider learning, especially abstract concepts in mathematics or philosophy, to be tedious or, worse, pointless. Even those of us who are excited by such abstractions find it difficult to make space in a busy day for learning. Dishes must be washed, kids must be fed, laundry must be folded, and the yard isn't going to mow itself; life is stingy about giving up time for intellectual pursuits. Making matters worse, humans tend not to even live a century on average, which is just about enough time to thoroughly explore one or two rooms in the vast library of knowledge.

Biological time constraints, coupled with the historical development of the scientific method and "specialization," have conspired to create a learning environment in universities and industry based around the concept of research. While scientific research, as a concept, might be the most powerful idea ever developed in human history, this environment leaves little room for cross-curricular study.

AI can access all information at all times, will not be distracted by what humans loosely call "life," and, most important, will evolve an understanding of knowledge by evolving to understand knowledge not by trying to painfully mold a poorly preadapted hunter-gatherer brain to the task of higher-order thinking but by evolving for the purpose of coming to a more complete understanding of data.

To repeat, I am not an AI specialist, but I have written extensively about the power of thought experiments and their importance for education. Even if AI fails to develop into a metacognitive theorist, the idea that it *might* do so is itself useful in the same way that the concept of time travel, even if impossible practically, is useful as a thought experiment for the purpose of clarifying intellectual models.

Having stated this, part I of this two-part book is about the old order of the history and philosophy of science. AI will be able to break this order down into its component parts and then build the history of philosophy and science back up in way that is not historically accurate to events that occurred but that is most useful for understanding scientific phenomena. Before a new order can emerge, however, the history and philosophy of science must be contemplated in its old order.

A

The Pre-Socratics

WHAT IS THE FUNDAMENTAL NATURE OF MATTER?

This question is the foundation of the human intellect. Eventually, attempts to answer the question would result in the development of the periodic table, which is the holy tablet of science and the peak artifact of the human intellect. Einstein, indeed, was not a scientist but a pre-Socratic unhinged from time; he stated that the fundamental nature of matter was energy rather than water, fire, air, or indivisible particles.

The historical connection between the pre-Socratic philosophical tradition, which was Greek and confined to the Ionian Sea, and the development of chemistry and theoretical physics can immediately be seen. There are no counterintuitive historical quirks here, no swerves. Western philosophy began with questions that are fundamentally understood now as "scientific" and did so in a way that Chinese, Indian, or American societies did not. If some Chinese or Indian thinker asked him- or herself, as the pre-Socratics did, "How many times can you cut a fish up before it ceases to be a fish and becomes a fundamental particle?," the question was lost to the historical record.

The term "pre-Socratics" might seem demeaning to these long-dead Greek thinkers, but they are probably largely literary creations (as was Socrates himself), a designation that has the benefit of making them mythically alive to readers rather than physically dead to history. Yet to call these philosophers "Ionians" would be to miss the importance of the perceived delineation between their scientific questions about the nature of matter and the questions that Socrates asked about the best way for a man to live.

But the distinction between the two types of philosophy might not be as important as the commonalities. The pre-Socratics and Socrates asked interesting questions, and questions seemed to be the way to attract students in a Greek society with no standardized forms of education. The ancient Greeks possessed no standardized body of medical, mathematical, or historical knowledge that needed to be mastered by students, and there were no mass-printed books or other forms of databases at the time that could hold such information.

In such an environment, education could not consist of forcing students to master a corpus of knowledge or a set of skills. Apprentice education must have been everywhere, from military training to grape farming, but such activities don't stimulate the mind or create conversation, so the teachers and the stage must have provided the only intellectual stimulation available in the Greek isles. The development of a set curriculum that required the mastery of a standardized set of skills and knowledge would not occur until the medieval period in Christendom with the creation of the *quadrivium* and the *trivium*. That standardization was more than 1,000 years away from the pre-Socratic era.

Thales, likely born around 625 BC (if he was a person at all), seems to have initiated the question about the nature of matter. Cut up any object into small enough pieces, Thales thought and taught, and you will find water. All matter therefore is a manifestation of water in different forms. His student Anaximander conceived instead of hypothetical particles that moved when sparked by energy. Anaximander connected this conceit to biology by stating that moving particles in the water and/or the mud could cause rudimentary life to form, which would then take different shapes.

While other pre-Socratics added theories about the nature of matter, it is Pythagoras (born around 570 BC) who developed a framework for the notion of a hypothetical world composed of pure mathematics. Pythagoras realized that the indivisible particles, conceived of by Anaximander and deemed as atoms by Democritus, could be thought of only as philosophical propositions. Treating them as such, a mathematician could simply compose a philosophical whole number and use that mathematical conceit to describe, say, triangles.

Pythagoras created a mathematical formula for right-angled triangles, which is to say that if one knows the length of two of the three sides of a right-angled triangle and one applies $A^2 + B^2 = C^2$ using a little bit of basic algebra ("algebra," as a term, did not exist then, but to avoid confusion, we'll go with it), then the third side can always be revealed. This works in a very practical way in the sense that one could use the Pythagorean theorem to sketch a plan in the sand, and then use ratios and proportions to scale the model up, and then actually build something.

However, Pythagoras also created the conditions of mathematical falsification. By stating that whole-number ratios are the fundamental essence of everything, he made a statement that could be disproved by mathematical logic. A right-angled isosceles triangle (a right angle with two equal sides) cannot be composed without using fractional units. The hypothetical world of mathematics worked through internal rules that seemed beyond the governing of the theorists, something that led many to believe that the mathematical world contained truths to be discovered rather than theories to be created.

B

Socrates, Plato, and Aristotle

IN THE INTRODUCTION TO *READINGS IN ANCIENT GREEK PHILOSOPHY, FROM Thales to Aristotle* (2016), Cohen, Curd, and Reeve note,

> Although we call these Presocratics "philosophers," they were, in fact, active in a tremendous number of fields. They would not have thought of astronomy, physics, practical engineering, and what we would call philosophy as separate disciplines, and they would not have thought that engaging in any of these would have precluded them from being active in politics. In a society that was more oral than literary, in which books were just beginning to be written and distributed, the Presocratics thought and wrote about an enormous number of things. In the ancient testimonies about the Presocratics, we find reports of books (or parts of books) on physics, ethics, astronomy, epistemology, religion, mathematics, farming, metaphysics, meteorology, geometry, politics, the mechanisms of sense perception, history, and even painting and travel. They wrote in poetry and they wrote in prose. They were as interested in the question of how we ought to live as they were in the problem of the basic material out of which the physical world is made. (5)

To restate a point, if anything connects the philosophical schools of Greece, it was the asking of interesting questions, and in an era before academic specialization, the questions could regard any form of knowledge. In addition, the presentation of the questions could be presented to students in various literary forms. This makes it complicated to delineate an actual historical record of the ancient Greek philosophers given that, while a person named Socrates may or may not have existed (he is referred to by Aristophanes in the play *The Clouds*, but Achilles was mentioned in a number of literary works outside of the *Iliad*, and this is not taken as proof of his physical existence) and did teach, the Socrates of Plato's dialogues might justifiably be considered a literary character based in history, similar to the Julius Caesar of Shakespeare's classic play. Or, in the same way that Achilles was a literary manifestation of Greek masculine virtues, Socrates may be the literary manifestation of Greek intellectual virtues.

For the purposes here, it's more efficient to view Socrates as a literary character and the great Platonic dialogues featuring Socrates as a literary and educational device for transporting Plato's theories about education and a life well-lived. Plato chose different mediums for conveying his philosophies, recognizing that a philosophy of life would best be transposed through stories involving a larger-than-life character, while a governmental theory about the necessity of philosopher-kings, or a realm of forms, was better stated in a drier manner.

To be plain, Socrates lives and acts as if he is a visitor from another realm, a perfect philosopher transposed into an imperfect world and therefore indifferent to what happens to him. Some historians have argued that the Athenians had good reason for putting Socrates on trial in 399 BC: the Athenians had just lost the Peloponnesian War, and Socrates spewed a philosophy that was friendly to the Spartans. His students had engaged in traitorous activities, and Athenian society no longer found anything cute in the countercultural activities of Socrates.

Whatever the historical conditions, Plato adopted them to show a philosopher-king placed on trial by an ochlocracy.

Understanding Plato's classic works on mathematics and the realm of (mathematical) forms helps us to understand Socrates as a character. Bertrand Russell, writing in a chapter titled "Retreat from Pythagoras" in *A History of Western Philosophy* (1946) on his discovery of mathematics and philosophy, recognized the influence of Pythagoras on Plato:

> The Pythagoreans had a peculiar form of mysticism which was bound up with mathematics. This form of mysticism greatly affected Plato

and had, I think, more influence upon him than is generally acknowledged. I had, for a time, a very similar outlook and found in the nature of mathematical logic, as I then supposed its nature to be, something profoundly satisfying in some important emotional respects. (227)

Although a fuller account of Russell's thinking will later be stated, he eventually (and probably unwillingly) came to understand, via the intervention of a ghostly Ludwig Wittgenstein, that mathematical reasoning consists of a series of tautologies. When Pythagoras creates a triangle, he creates a theoretical situation that works by its own invented logic. In that respect, the Pythagorean theorem operates like the internal rules of grammar, where a grammatical sentence should not begin with a preposition because, by definition, grammatical sentences do not begin with prepositions.

But it is the difference between mathematical tautology and linguistic tautology that makes it more difficult to see mathematics as human-made circular reasoning and seems to give mathematics a greater aura of the mystical. If a Greek philosopher took a stick and drew a triangle in the sand and then worked out the measurements with the Pythagorean theorem, it would be apparent to everyone that the shape in the dirt was imperfect, yet the concepts involving the lengths of the sides would seem to be intellectually conveyed as perfect measurements. This led to the belief, taught by Plato, that a perfect triangle must exist somewhere.

Where did the perfect triangle exist? In the *realm of forms*. The triangle in the dirt is an imperfect shade of the perfect triangle that exists in the realm of forms, but the act of analyzing that triangle in the dirt allows humans to access, via mathematics, this perfect realm.

For scientists, engineers, and computer programmers who develop mathematical models and then use those models to send rockets into space or to construct bridges, the idea that they are pulling perfection from a Platonic realm and trying to create objects in this shadow realm that are as close as possible to the perfect is probably appealing and leads to the notion that there are existing theories and rules waiting to be "discovered." That conceit, as much as anything, generates the idea, oddly prevalent, that the universe must be created by some intelligence.

To that end, Plato's student Aristotle argued against Plato's realm of forms. When dealing with quasi-mystical ideas like the perfect shape, Plato makes a certain kind of sense, but the notion that every rock, tree, or soda can in the human world is an imperfect shade of a rock, tree, or soda can in a realm of forms seems to stretch to the human intellect a little too far away from reality.

Aristotle argued for materialism, the idea that objects must be made of some kind of material if those objects can be said to exist. (The concept of an object that gives off no sensory data that would allow a human to detect it, an "atomic fact," would in the early 20th century fascinate Wittgenstein.) In some ways, Aristotle intuited that the modern concept of the "supernatural" is, as it always has been, complete nonsense.

After all, if someone sees a ghost, then that ghost must be made of some material that can reflect light, or else the ghost is phosphorescent and creating its own light, but that is no more supernatural than what a firefly accomplishes every summer night. Poltergeists who slam doors must have some atomic structure, or else they could not interact with another atomic structure. If the poltergeist interacts with a door, then at some point, a measurable force is being applied, which would make the interaction no more mysterious than gravity acting on a coffee cup or the wind shaking the top of a tree.

Aristotle, of course, created the syllogism. The syllogism can be thought of as a verbalized tautology where a definition is created and then reaffirms itself. For the purposes of this book, it is Aristotle's conceit of motion that is important. The two significant passages from books 10 and 12 of his *Metaphysics* are as follows:

> Now if there is something that is capable of initiating motion or of acting, but it does not actually do so, there will be no motion; for what has a potentiality need not actualize it. It will be no use, then, to assume everlasting substances, as believers in Forms do, unless these include some principle capable of initiating change. And even this, or some other type of substance besides the Forms, is not sufficient; for if it does not actualize its potentiality, but its essences is potentiality; for there will be no everlasting motion, since what has a potentiality need not actualize it. There must, then, be a principle of the sort whose essences

is actuality. Further, these substances must be without matter; for they must be everlasting if anything else is to be everlasting, and hence they must be actuality.

Now a puzzle arises. For it seems that whatever actualizes a potentiality must have it, but not everything that has a potentiality also actualizes it; and so potentiality is prior. But now, if this is so, nothing that exists will exist, since things can have potentiality to exist without actualizing it. And yet, if those who have written about the gods are right to say that all things were together, the same impossibility results. For how will things be moved if there is no cause [initiating motion] in actuality? For surely matter will not initiate motion in itself.

Here Aristotle frames the notion of cause and effect with motion, a natural intellectual progression given the limitations of the human life span. If someone pries a rock free from a mountainside and then watches the rock fall down the mountain, one can watch a chain of events set in motion. Eventually, one can reason backward to a firm logical conclusion:

If, then, the same things always exist in a cycle, something must always remain actually operating in the same way. And if there is to be coming to be and perishing, then there must be something else that always actually operates, in one way at one time and in another way at another time. This [second mover], then, must actually operate in one way because of itself, and hence either because of some third mover or because of the first move. Hence it must be because of the first mover; for the first mover will cause the motion of both the second and the third. Then surely it is better if the first mover is the cause. For we have seen that it is the cause of what is always the same, and a second mover is the cause of what is different at different times. Clearly both together cause this everlasting succession. Then surely this is also how the motions occur. Why, then, do we need to search for any other principles?

This argument, of the necessity of a "prime mover" or first cause, is generated from the notion, based on a partial understanding of motion, that some big movement causes a smaller secondary movement and so on. Shove a big ball so that it rolls into a smaller second ball, and that smaller second ball will run into a smaller third ball and so on, and you will see the force of motion dissipate. The prime mover must be the strongest force in this chain, a strength worthy of worship, even if it did not actually care about the human lives that the first cause of motion eventually created.

Pythagoras and Plato created an ethereal conceit of mathematics that would prove to be useful delusion for future mathematicians who sought to superimpose their theories from the realm of forms into the realm of actual human existence (Plato's shadow realm). Aristotle constructed a theory of cause and effect built on motion, and this was based on the analogy of causal motion based on the limited sensory experience derived from watching motion on Earth through a limited time frame of observation.

Neither Aristotle nor Plato observed or even conceived that all motion involves the loss of heat.

C

The Hellenistic Era

In *A History of Western Philosophy* (1946), Bertrand Russell, in a chapter titled "Currents of Thought in the Nineteenth Century," wrote,

> Until about the end of the eighteenth century, scientific technique, as opposed to scientific doctrines, had no important effect upon opinion. It was only with the rise of industrialism that technique began to affect men's thought. And even then, for a long time, the effect was more or less indirect. Men who produce philosophical theories are, as a rule, brought into very little contact with machinery. (275)

Pythagoras, Plato, and Aristotle split mathematics from the "real-world" forces of motion and cause and effect. These forms of philosophy could touch and interact, but they would remain largely separate from each other until the machine age generated questions about mathematical applications that would eventually result in the restructuring of mathematics.

It is important to state these facts here because, eventually, the study of actual machinery would lead to an understanding of heat loss (entropy) and its universality, but this did not happen until the 19th century. The Hellenistic era of Greek philosophy began roughly after the death of Alexander the Great (356–323 BC) and is considered to have continued on until about 31 BC, although the Roman Republic's conquest of Syracuse in 212 BC is often considered to have at least diminished, if not destroyed, this era of Greek philosophy.

A sophisticated concept of motion did develop in the mind of at least one Hellene, Aristarchus, who conceived, at some point between 127 BC and 151 BC, of a "heliocentric" solar system. Aristarchus, like Euclid, could draw triangles or circles on just about any surface and work out accurate measurements and calculations by doing so. But again, the study of the night sky seemed like a separate endeavor from the study of machines, as if the stars moved by different rules of motion and mathematics from terrestrial machines. Was this false dichotomy (not really proved as such until Newton) a result of Plato's conceit that the realm of forms was "out there" somewhere and that objects in the night sky were closer to the realm of forms than to the earthly shadow realm?

This is just speculation of course, but something kept Aristarchus and his philosophical concept of the solar system from interacting with the machine work created by the Hellenic engineers. The Greek inventor Philo of Byzantium (280–220 BC) supposedly developed machines that could send ciphers and created air pumps that worked with chains. Heating up water, he found, produced steam that could be compressed and used to move machines. Steam power was used at a rudimentary level and only for entertainment, but no one seemed to note the immense philosophical importance that can be derived from understanding that hot steam cools down and loses its force, always, rather than vice versa.

If anyone could have derived scientific principles from the study of machinery, it would have been Archimedes (287–212 BC), but he did not, and it is a puzzle how a man of his titanic intellect somehow managed to miss the universality of entropy in the motions of machines. Archimedes, after all, created scientific notation as a means of keeping track of large numbers (he imagined all of the grains of sand in the world, hence his nickname the "sand reckoner") and also generated a theory of machine motion from an understanding of how the length of the fulcrum affected the ratio of the weights.

Most interestingly, he theorized that objects have a center of gravity. It should be noted that this understanding of weight is temporal and understandable within the scope of human life span. A hundred generations of scientists could discover essentially the same center of gravity from a large stone, but even if they kept careful records, none of them could derive from observation that the stone was (as it is) radiating away.

Archimedes could have used gravity as a continual force in the development of his machines while also recognizing that muscle force, or energy captured from running water, dissipated. The question as to why gravity seemed permanent, but the outside forces seemed temporal seems not to have arisen. This may be because Aristotle had framed the notion of force in terms of cause and effect, with all observable physical actions in the world considered to be continuations of the prime movement. Yet Archimedes, for all his genius, could no more conceive of the question "Why does everything decay?" than a 16th-century Confucian scholar could conceive of asking, "Why is there no Chinese literary tradition of romantic rebellion against authority?"

Syracuse, the island home of Archimedes, sided with the Carthaginians in the Second Punic War. The Greeks, understandably alarmed at the aggressively expansionist Romans, lost their independence to Roman conquest. The Romans valued engineering because this created both machines of conquest and the means of maintaining control, through roads and aqueducts, after their enemies had succumbed. Yet the Roman numeral system occluded the Roman mind to any form of mathematical thought. The Platonic realm of perfect mathematical forms disconnected itself from Western civilization and would float in silent isolation until the time of Fibonacci some 15 centuries later.

D

Rome and Lucretius

THE FACT THAT THE ROMANS WERE SUCCESSFUL ENGINEERS BUT NOT necessarily mathematicians actually exposes the limited practical scope of mathematical reasoning. Practically speaking, it's possible to build just about anything with just uniform measuring tools and the kind of dead reckoning that becomes second nature to workers. For example, someone looking to mulch a circular area around a tree probably will not measure the radius of the circle and then use a geometric formula to determine the specific area of the circle. Instead, she will probably guess that four bags of mulch will be needed and then buy five bags. If some mulch is left over, it can be saved for later.

The Egyptians, Romans, Aztecs, and Mayans all built grand structures without any kind of theoretical mathematics. The structure of a pyramid can be derived by children playing with blocks. Likewise, when cannon technology was developed, none of the rough soldiers who aimed the cannons at castle walls knew anything about Aristotle or trigonometry. Early cannoneers did not understand that a cannonball will always reach its farthest distance at a 45-degree angle. Instead, they played with gunpowder packing and cannon angles until they developed a highly honed form of battlefield intuition. Complex forms of mathematics were not—and are not—necessary in most forms of useful applications.

The Roman Republic seemed not to act as a muse for the poets and philosophers. The treachery of Caesar and Antony inflamed the pen and inspired the speeches of Cicero—but only as vitriol. When the republic collapsed and the need for a national epic became apparent, Virgil generated the *Aeneid*. Philosophy continued on in the form of stoic traditions, with Marcus Aurelius eventually reminding his fellow Romans to keep death on their minds at all times because morbidity reminds men of how temporary everyday concerns over status and relationships can be.

But prior to all of this, in the last days of the republic, just a few years before Caesar grabbed control of Rome and the senatorial assassins then grabbed his toga, the great philosopher Lucretius (99–55 BC) wrote his *De Rerum Natura*, or *On the Nature of Things*. The term "essay" (meaning "attempts") would not be invented until Montaigne in the 16th century, and neither a speech nor a play would have been a fit medium for the ideas of Lucretius. He chose to write an epic poem, but his words read very much like a paper of scientific skepticism.

On the Nature of Things comes to us in six books, and to try to encapsulate a work of such breadth and genius would be to violate it, but passages will be chosen here that will give the reader some idea of how Lucretius was able to intuit some of the most impressive concepts of 20th-century physics. It's implicit in the second law of thermodynamics that time travel cannot occur, but if it were possible, then Lucretius would be the best case for a time traveler, as he seems to foresee many of the great theoretical breakthroughs that would not occur for 20 centuries:

On the supernatural (from Book I: Matter and Void):

> This dread, these shadows of the mind, must thus be swept away / Not by rays of the sun nor by the brilliant beams of day, / But by observing Nature and her laws. / And this will lay / The warp out for us—her first principle: *that nothing's brought Forth by any supernatural power out of naught.* / For certainly all men are in the clutches of a dread—/ Beholding many things take place in heaven overhead / Or here on earth whose causes they can't fathom, they assign / The explanation fort these happenings to powers divine. / Nothing can be made from nothing—once we see that's so, / Already we are on the way to what we want to know. (7)

On the notion that the elemental base of everything is fire and a little bit more regarding vague authorial styles (from Book I: Matter and Void):

Therefore those who think the basic element is fire, / And it is out of this material that the entire Universe is made, have clearly wandered far astray / From the path of true reasons. The first warrior in the fray is Heraclitus, a leading light for the murkiness of his style / More so among the shallow, than those Greeks who find worthwhile / The search for truth. For idiots admire things all the more When they discern them hidden in tangled words. (21)

On the four elements of earth, air, wind, and fire (from Book I: Matter and Void):

In short, if the whole universe were fashioned from these four, / And everything dissolved back to these elements once more, / Why are they called the elements of things? Why not instead / Consider things their elements, and turn it on its head? For they are born of one another, ever changing hue and / Changing their entire natures with one another too. / But if by chance you think that particles of earth and fire / And windy air and flowing water so combine together / That their natures do not alter in their unions, you will find / That nothing can be fashioned from such elements combined, / No animal, nor anything inanimate, like a tree, but every item added to this jumbled assembly will display its own nature. (25)

Most tellingly, from Book II: The Dance of Atoms:

All bodies of matter are in motion. To understand this best, / Remember that the atoms do not have a place to rest, / And there's no bottom to the universe, since Space does not / Have limits, but is endless. As I have already taught and proved with reason irrefutable, it opens wide / And far in all directions, measureless on every side. / And therefore it is obvious no respite's ever give / To atoms through the fathomless but, rather, they are driven / By sundry restless motions. After colliding, some will leap / Great intervals apart, while others harried by blows will keep / In a narrow space. Those atoms that are bound together tight, / When they collide with something, their recoil is only slight / Since they are tangled up in their own intricate formations:

Such are the particles that form the sturdy roots of stone, / And make up savage iron and other substances of this kind. / Of the other particles drifting through the vast deep, we find / A few leap far apart and bounce a long way back again, Providing us with thing air and the shining of the sun. / And many more besides stray through the void, either out cast / From combinations, or which alliances could not hold fast / In harmonious motions. (39)

Although Lucretius ridiculed the elemental conceits of the pre-Socratics, his questions about the nature of the universe were framed by their intellectual creations. It was the process of thinking about the false conjecture of using air, wind, fire, or water as the fundamental unit of all matter that led Lucretius to the conceit of atoms in "sundry, restless motions."

On the Nature of Things effectively renders the concept of "supernatural" absurd. Try to define the term, and you will quickly see that there is no actual definition, just improbable explanations, mystified with vagaries, for ordinary science. For humans to understand anything, there must be sensory data (something to taste, touch, hear, see, or smell), and if sensory data exist, then the object is natural and not supernatural. Seeing a spirit, smelling brimstone, or hearing a bump in the night are all ways of interacting with the natural world.

Lucretius indicated that atoms in motion, their compilations creating matter, their motions changing forms, simply *exist* because that is the natural state of these things. No gods create or interfere; the very brains that conjure up these deities are made themselves of ever-moving atoms. Such a prophetic poem leaves the reader with a simple question: Did Lucretius foresee the future, or did he create it? Is the scientific image he developed in his poem an intuition of a future model developed through research by scientists, or did Lucretius create a framework that caused future scientists to conceptualize their findings in terms of indivisible particles in motion?

The question is crucial because for all his brilliance, the concept of universal decay did not radiate from the mind or the pen of Lucretius.

E

The World Historical Narrative

In the early fourth century, the emperor Constantine synthesized Roman power with a nascent and nebulous form of Christianity. This synthesis re-created Western civilization in a number of ways, but philosophically, the most important changes came in the development of two overpowering intellectual tensions. The first tension was between Christian teachings and the actual necessities of action inherent in the practice of political power. Eventually, this led to the development of new antithetical philosophical systems that reveled in rather than refuted the aggressive use of state power. The second tension involved using Greek logic and eventually materialistic logic to justify a belief in the tenets of Christianity.

The existence of the first tension is what made Christendom cum Western civilization unique among early empires. With the development of hierarchical political structures, ruling elites found it convenient to "choose" religions and philosophies that upheld the existing structure by inculcating in the people a sense of ethics that served the interests of the state.

This could happen in a variety of ways. In India, the conquering Aryans imposed Hinduism, complete with a stay-in-your-place ethical structure in the form of the caste, that evolved to solidify class, racial (sometimes), and gender distinctions. In China, Emperor Wu Di (156–88 BC) recognized that the centuries-old philosophy of Confucianism, with its emphasis on respect for male authority, would serve the patriarchy well at the level of both the home and the empire.

When Islam developed after the Arabs conquered sections of the Byzantine and Persian empires, the character of Muhammad and the religion of Islam were either created entirely or adapted partially by the Arabic conquerors who sought to create an empire and a patriarchal state. Nothing in the texts of Islam, for example, would condemn a man with political aspirations from slaughtering non-Muslims in war and taking their women as slaves. Something similar happened in Mesoamerica, where the Aztecs justified and upheld their empire by sacrificing conquered tribes to the sun god.

Christianity, however, fit less well with the purposes of the state. Jesus, like Yahweh, is a literary character and probably a composite character at that. Jesus contained elements of Socrates in that he is a male teacher with male followers who goes willingly to his death when condemned by the state (a cup is involved in both stories). Although four "synoptic" (meaning "see together") Gospels would eventually be chosen for the New Testament, historians now identify at least two dozen narratives regarding the life of Jesus. All present different (sometimes radically different) narratives about the life and teaching of Jesus, making it impossible to construct a central story line about his life.

Before moving on, a point needs to be made about the structure of the four synoptic Gospels: Matthew, Mark, Luke, and John. All four come in narrative form and are therefore connected to a literary structure used in Genesis and the Dialogues of Plato. Neither the Confucian Analects, the Hindu oral traditions, nor the Koran contain narrative biographical structures. A Muslim believer is not invited, for example, to "know" Muhammad in the same way that a Christian believer might be asked to do so with Jesus.

Yet in the murkiness of the Gospels, it was still possible to read some core central theology, and it was not one that was friendly to the exercising of state power. Jesus was condemned and then crucified by the state, not made emperor. Furthermore, Jesus specifically blessed peacemakers and those with compassion for the poor and the hungry.

Nietzsche read Jesus' condemnations of the elites as the whining of a loser, and many readers of the Gospels see Jesus' apocalyptic predictions as evidence of an elaborate revenge fantasy. But Christianity could not exist as a theology without "faith" in some future where the power structure would be upended. What else were a group of monotheists who were subjected to occasional state prosecution in Rome supposed to believe in?

The point is that when Constantine elevated Christianity to what would eventually become the official religion of the Roman Empire, he probably did not put that much thought into the decision. Certainly, Constantine believed that a monotheistic religion would provide more cultural unity than polytheism, and Rome needed the cultural unity that a singular religion could provide. Constantine thought that he was adopting a religion where the concept of "one god, one emperor" would uphold the state and that he could create a hierarchical structure of bishops that would place Rome's culture beside an official power structure that would rule Rome's body politic.

The bishops solved the problem of Christianity's nebulous theology by creating a hierarchical structure of interpreters who would compile Judeo-Christian texts and then interpret them for meaning. By doing so, the catholic church (the small "C" in "catholic" is deliberate, as before October 31, 1517, the word was an adjective, not a proper noun) assumed the same kind of power, based on circular reasoning, seen in China, India, and eventually the Islamic world and Mesoamerica. The power was not in the unreadable texts but in the structure that would interpret those texts.

Christian theologians sought to create a singular narrative for Christian believers, and this was supposed to be the only viable narrative, with all others descending into heresy. Although the Bible is an anthology of early Jewish and Christian texts and was not compiled until the fifth century, the compilation of the texts contained a deliberate narrative thread that, roughly speaking, can be characterized as follows:

> God created the universe in seven days. Humans were placed in Paradise, and through their sin, made a perfect world into a fallen world. God's people were the Jews until the arrival of Jesus, and his life and sacrifice offered humans salvation. Upon his death, the disciples of Jesus carried his message on through the Church, and those who joined the Church were promised eternal salvation after the Rapture and Armageddon. Those who refused the salvation would be condemned eternally to Hell.

Other religions contained narratives about creation and destruction, but Christianity sought to impose a single narrative on the history of the universe and to construct a narrative around the importance of humanity. The scientific enterprise would, eventually, generate evidence that the Christian narrative could not explain, but the desire for a singular explanation and the notion that it exists to be found would continue, shaping the Western mind in the same way that a drinking vessel will force a new fluid to adapt to the same parameters as the previous fluid.

F

The Theologians

CHRISTIAN THEOLOGY IS NOT THE EQUIVALENT OF ALCHEMY; IT IS NOT a patently wrongheaded endeavor that gave way to a science in the way that alchemy evolved into chemistry. Christian theology, instead, can be seen as the use of logic for the purpose of trying to prove an already existing belief system. To that end, the theologians examined the various arguments in favor of a belief in God and of the Christian tenets and then, using analogy, classified a certain set of poor logical justifications as "fallacies." Understanding and applying the fallacies is as complex an act as understanding and applying mathematical formulas because both require an assessment of an original situation, the determination of what fallacy or formula needs to be employed, and then an application to the given scenario or problem.

The Christian narrative became a powerful tool for free thought and for the development of logic and science precisely because it provided something to argue against. The great theologians, Thomas Aquinas (1225–1274) among them, made it intellectually and spiritually respectable to develop arguments for the faith against a hypothetical atheist. It might be possible to imagine the theologians constructing their defensive arguments against rhetorical attacks or questions from Jews or Muslims, but the trouble with any such interfaith discussion/debate is that any logical question that exposes the fallacy of one faith exposes them all.

For example, a Muslim who sought to explain the superiority of Islam over Christianity could do so by stating the Koranic story about Jesus. This story holds that Jesus was a great prophet but was not executed (an impostor took the nails for him). The Muslim would have to argue for the veracity of this narrative by stating that the Koran is more reliable than the New Testament. Since no proof exists for the veracity of either the New Testament or the Koran, the Muslim simply reverts to a reliance on faith, and that faith could be proven only by invoking some conceit unrelated to logic, such as Muslim military and/or financial dominance.

Initially, after the time of Constantine, Christian theology was defined by St. Augustine (AD 354–430) who recognized the seeming contradiction between the concepts of free will and of an omniscient god. Augustine answered the paradox by siding with the omniscient god and doing away with the conceit of free will. Free will, however, turned out to be core to the business model of the catholic church (there's not much for the priests and nuns to do if everything is predetermined), so Augustine's theology would become idle until after the Reformation, when Jean Calvin incorporated predetermination into his conceit of "the elect" of those who were preordained for heaven.

We can now see that St. Augustine engaged in the same kind of logical fallacy, the tautology, that mathematics is based on. Tautologies create false logical premises that, under hypothetical conditions, result in paradoxes. The only way to resolve the paradox is to question some aspect of the tautology. An omniscient power, for example, does not exist, and this renders questions about its hypothetical limitations moot. Infinity, also, does not exist, and this renders the many mathematical paradoxes inherent from the term moot as well.

After the death of Augustine and the collapse of Roman authority (events that were roughly concurrent), the catholic church held the societal structure of Western Europe together in the absence of secular Roman institutions. By the sixth century, the educational structure was almost entirely self-contained within the Church and consisted of the seven liberal arts with the *trivium* (grammar, logic, and rhetoric) forming the lower portion, taught first, and the *quadrivium* (arithmetic, geometry, astronomy, and music) the higher portion.

This basic curriculum contained within it the propositions for an intellectual revolution, but it had to be synthesized with Aristotelian reasoning.

Prior to the 21st century, when historians seemed universally seized by the desire to decenter Western civilization from the world historical narrative, the Arabic era of philosophy was seen (particularly by Bertrand Russell) as being a period where the Greek classics were translated and kept in storage. This is largely true. By the time of the second great caliphate, the Abbasid (750–1258), was established, Aristotelian philosophers, facing the perplexed wrath of Justinian (482–565), had been expelled from the Byzantine Empire for two centuries. The new Arabic/Islamic state was less hostile to Aristotle than had been Justinian, probably because the new Arab elites developed political power and made fortunes by relying on the engineering and bureaucratic expertise of the people whom they had conquered, so a general hostility to Greek philosophy was not a part of their intellectual makeup.

Also, the early caliphs of the Abbasid Caliphate read the Koran as being an invitation to learn, with higher thought being a form of worship. In this environment, a House of Wisdom, or library cum university, developed in Baghdad in the ninth century. Here, Aristotle's philosophies were translated into Arabic and commented on, chiefly by Averroes (1126–1198), who explained Aristotle's philosophies with a patient and systemic form of logic that would enliven the minds of Christian theologians.

Islam functions primarily as a political and legal doctrine, not an intellectual discipline, and so free thought and philosophy flourished for a short while until the more doctrinaire authorities got smart enough to realize what they should be afraid of. Just as the Islamic religious authorities shut the book on Aristotle, the monks of Christendom were opening it.

Between the fifth-century collapse of Rome and the development of the Holy Roman Empire in 800, Christendom lacked for libraries, and, Boethius aside, philosophy stagnated. The Carolingian Renaissance, begun by Charlemagne (747–814), swirled the intellectual waters again by routing books and manuscripts from far-off lands and private collections into the monasteries, where monks copied the works by hand.

These monasteries turned scriptoriums were staffed by well-educated theologians, many of whom had mastered the *trivium* and the *quadrivium*, so the arrival of a radically new form of thought (materialism) practiced by an ancient Greek master, the antithesis of Plato, caused an excitement among the theologians that was strong enough to overcome the trepidations of the church bureaucrats.

The traditional narrative teaches that French theologians, especially Peter Abelard (1079–1142), tried to immediately determine whether Aristotle's form of materialism could "prove" the already existing tenets of Christian faith. When this caused an intellectual disruption (the answer, as Abelard noted, could be "no," which left it an open question as to whether reason could disprove faith or vice versa), the church first banned the teaching of Aristotle and then, when that proved impossible to enforce in a climate where students paid teachers in an open market, decided to embrace Aristotle's reasoning.

St. Thomas Aquinas, a Newtonian intellect, synthesized Aristotle's reasoning with catholic dogma, as the traditional story goes, and for the next three centuries (until Galileo), the Scholastics would, by rote, pass on and defend the Aquinas/Aristotle synthesis. This synthesis, where Aristotle reigned infallible, included the belief in a geocentric planetary system and would eventually be challenged by Copernicus and Galileo, while the Scholastics (Latin for "schoolmen") would try to defend the center of their system against the plain scientific facts in favor of a heliocentric "solar" system.

Yet, as Jennifer Michael Hecht writes in *Doubt: A History* (2004) (a lengthy passage but crucial for understanding the effect that Aristotle's philosophy had on the grammar of the *trivium*),

> The most important of all the new tests was Aristotle's *Sophistic Refutations*. It was a study of fallacy: how words work and what to call the ways in which they can be deceiving. It appeared in Latin about 1120 and just plain took over. For one thing, it was easier to understand than the others, and for another, it offered instruction about work that was yet to be done and how to do it. The Schoolmen picked it up with glee. They started to take apart the language to see how it worked so they could be sure of speaking the truth—there were treatises, for instance, on *syncategoremata*, words that cannot be subjects (such as: *every, because, and, or, if*), which made logicians suspicious of sentences with such words

in them. In fact, the way that Latin words worked soon became more important than their content. What is more, although word order is unimportant in Latin, the Schoolmen assigned meanings to Latin word order and thus expanded their project. That is, the Schoolmen learned from Aristotle that this conceptual game of logic could be used to negotiate fallacies and they turned Latin into a symbolic logic system, rather than a system of communication. The Schoolmen thus used purposefully silly content, so that content would not distract them. That is why it has seemed like such hocus-pocus to the later Europeans.

Apparently a sentence was constructed so that one could argue that it was true and then argue that it was false, and the idea was that one would learn something, or hone one's truth-finding skills, by analyzing these claims. Scholastic authors liked citing famous ancients in their content-nonsense texts. Consider some of their sentences:

Varro, though he is not a man, is not a man, because Cicero is not Varro. A head no man has, but no man lacks a head. Socrates is whiter than Plato begins to be white. Socrates will as quickly have been destroyed as he will have been generated. A horse is an ass. God is not. No man lies. Some horse does not exist. What Plato is saying is false. Socrates wants to eat.

A variety of exercises and games were being worked on here. Sophisms were one major project of these weird statements. Sophisms were arranged in themes. That is, they were understood to exemplify, for example, themes of signification, supposition, connotation, and the insoluble. The sophisms "This dog is your father" is true and intelligible because the dog is yours and the dog is a father; you own this father; this dog is your father. The best here are the insoluble, the most famous of all being: "What I am saying is false." (254–55)

If we accept the proposition that the *trivium*, with its focus on grammar, must have prepared the minds of the Schoolmen for an understanding of Aristotle's complex theses on grammar (as if students of arithmetic were suddenly handed a textbook on algebra or programmers working on a DOS system found the opportunity to work with Word), then Hecht argues that ridiculing the Schoolmen is a way of ignoring their contributions. (The "dunce" cap, recall, is named after Duns Scotus [1265–1308], who had mastered the art of what Hecht calls "weird statements.")

Grammatical questions such as those offered by the Schoolmen offer important epistemological lessons, ones that would be lost when scientists began to disdain any nonmathematical argument in physics as being the only kind of argument that people who can't do the mathematics could understand.

Ridiculing and ignoring theology/Scholasticism as an intellectual source has led to what should be some embarrassing misappropriations in the intellectual history of mathematics. For example, the great mathematician Georg Cantor is credited with recognizing an infinity paradox in 1874. If there is a one-to-one ratio between the numbers 1, 2, and 3 and 21, 22, and 23 and the numbers go on forever, then there must be no actual difference between the outcomes when adding the numbers of both lines. Put another way, there can be infinite decimal points between the numbers 1 and 2, just as there are an infinity of numbers. Cantor theorized that this must mean that there are infinities of different sizes.

Yet here is the medieval theologian Robert Grosseteste (1168–1253) writing nearly six centuries before Cantor:

> But it is possible that an infinite sum of numbers be related to an infinite sum (of numbers) in every numeric ratio and also in every non-numeric ratio. And there are infinities which are greater (*plura*) than other infinities, and infinities which are smaller (*pauciora*) than other infinities. For the sum of all numbers both even and odd is infinite. For (the sum of the even and odd) exceeds the sum of the even by the sum of all odd numbers. Moreover, the sum of the numbers double continuously from unity is infinite; and similarly the sum of all the halves corresponding to these doubles is infinite. And the sum of these halves is necessarily half the sum of their doubles. Similarly, the sum of all the thirds corresponding to the triples. And the same thing is clear in all species of numeric ratio, since the infinite can be proportioned to the infinite in any of these ratios. (475–76)

If Grosseteste, who would not even have had ready access to the Arabic numerals, could intuit what became (unjustly) known as Cantor's theorem, then the Christian theologians deserve to be taken seriously as thinkers. As Hecht indicates, what may be seen as nonsense words were in fact important epistemological exercises of the kind that mathematicians should engage in as part of their education.

Try engaging in a more modern version of the Schoolmen's grammatical exercise. Find one way that the following sentence can be true and one way it can be false: *Veterinarians are people doctors.* Someone attempting to make this statement true might inculcate into the words a subtextual meaning: that by helping sick pets, veterinarians are also helping people. One could read the same subtext into the statement *pediatricians are parent doctors.* Both statements would be strengthened and their meanings made more immediately accessible by adding the word "too" at the end.

One could also use the surface-level meaning of the statement to declare it to be true. Veterinarians treat animals. People are animals; therefore, veterinarians treat people. This is logically correct in a syllogistic sense but treats the meaning of the words using formal definitions rather than the colloquial meaning.

To declare that "veterinarians are people doctors" is false, one would simply need to stay at the most surface-level version of the statement. Veterinarians, by definition, are animal (in this context, meaning all animals but humans) doctors and therefore not people doctors.

Let's do the same activity with the most basic of mathematical equations: $2 + 2 = 4$. To proclaim this as true is to categorize numbers as unalterable definitions. A truth can be declared, such as A. All men are named Socrates, B. This is a man, C. He must be named Socrates. In this sense, $2 + 2 = 4$ is always true. But if you had two women in a room, one of whom was 20 weeks' pregnant and another who was holding a newborn baby, and you asked someone in the hallway how many "people" were in the room, the response would differ based on the definition of "people." Some would declare the fetus a person, and others would not. The number categories at that point become, practically speaking, hostage to reality.

Think of a way that $2 + 2 = 5$ could be true. To do this, one could establish that five diamonds weighed as much as four gold bars. If the number to the right of the equals sign is 4 in the first case but 5 in the second case but the weights of both are the same, then 4 = 5. If 2 + 2 = 4 and 4 is the same as 5, then 2 + 2 = 5 in that context. Mathematicians might consider this a cheat in the same way that English majors consider Ebonics or Esperanto to be fake languages, but it is a legitimate proof.

Paradoxes always contain some tautology that does not actually exist, but is a problem of a hypothetical definition. To say that God is omnipotent but that there is also free will would seem to be an unsolvable contradiction, but an omnipotent god does not exist, so neither does the paradox. Neither Grosseteste nor Cantor solved the problem of infinity because infinity, also, fails to exist.

G

Gunpowder, the Compass, the Clock, and Block Printing

To stay aligned with the promise made in this book's introduction, a "this happened and then this happened" traditional history of science won't be presented here in part I. Well-written traditional histories of the objects and concepts can be explored in the references. The purpose of this section is to consider philosophical changes made.

The story has it that, in the early 13th century, a militarily gifted Mongol named Temujin united the horsemen and horsewomen of the steppes north of China, and then he and his successors conquered a land empire large enough to connect China with Russia and some parts of Central Europe. The late medieval period, or High Middle Ages, in Europe was marked historically by the arrival of technologies and viruses from the East that came as a result of these Mongol-created trade routes.

Genghis Khan probably never existed; it's more likely that he was invented (decades after the initial Mongol conquests) by whoever wrote *The Secret History of the Mongols*. Genghis can be only a literary figure and most likely is a composite character intended to display the ruthless Mongol military intelligence that allowed them to create such an impressive empire. The Khan was conjured into existence by newly emergent Mongol rulers who needed a simple history, focused on a legendary figure, that would send a clear message to the newly ruled. In this, the Mongols likely copied from the Muslim rulers whom they frequently encountered and conquered.

The reason for bringing this up here is to note that, until the arrival of the printing press, history as it is currently conceived did not exist. The printing press, a synthesis of Chinese block printing, German metallurgy, and the phonetic alphabets used by German- and English-speaking peoples, evolved in the German provinces in the middle of the 15th century. Literacy evolved in a symbiotic relationship with the press, in the same way that the skill of typing evolved symbiotically with the widespread use of computers in the 20th and 21st centuries.

Literacy and the press created more effective forms of record keeping and, at the same time, more demand for what now might be called "news." This meant that historical records were kept in greater numbers and closer to the actual events that occurred. The printing press created the conditions by which both modern concepts of journalism and modern concepts of history could be created.

Martin Luther (1483–1546) might be the first historical figure in the sense that he lived in a time and place where literacy and the printing press were more or less endemic, and Luther seems to have written just about everything he ever thought, giving modern historians a thorough vantage of his life.

The printing press did not create modern science, as the scientific method existed to some degree in the Indian and Arabic worlds. However, when nascent variations of the scientific method did develop in India and in the Abbasid Caliphate, those ideas did not "stick" in society because scientific discussions remained limited to a small number of intellectuals. In India, we are not even entirely sure how the numbers 0 to 9 evolved, although they almost certainly are connected to Hindu religious practices, which frequently included meditations and ruminations on the concepts of infinity and the void, necessary components for contemplating the immense periods required for a soul to reach Nirvana.

Eastern "religions" (the quotations indicate that the term is not very well defined and is often ill-used) tended to focus on the development of a state of mind rather than on salvation. The contemplation of esoteric questions, a technique of Chinese Taoism, forces the thinker out of everyday concerns and into a state of peaceful contemplation. The same is true of meditations in Hinduism, and if the purpose of a religious practice is to reach a state of mind, then philosophical or mathematical questions, by definition of little practical purpose, are effective for achieving or finding that realm of the mind.

It is not enough for ideas to just develop; they must be recorded and taught if they are to permeate a culture. It may also be the case that those ideas must, in a Hegelian sense, clash against some existing synthesis or dogma. The friction makes the fire. For example, Hinduism, with its notion of the soul changing over long periods of time and with its vast speciation of gods, might make India seem like the natural place for a theory of evolution to emerge. Yet when Darwin developed evolutionary theory, he did so by thinking about the problems with the creationist "watchmaker" analogy, not by contemplating the infinite and the void. The niggling feeling that one gets when encountering a theory or explanation that seems not quite right seems to be the impetus for creative thought.

The compass and gunpowder, both Chinese technologies, seem to have been originally used by mystics. The Chinese famously used gunpowder for fireworks or stuffed sections of bamboo with it to make firecrackers. The sudden explosion, it turned out, was useful for scaring enemy horses. Eventually, the Mongols brought gunpowder to Europe, where European metal casters found that they could easily modify church bells into cannons.

The compass itself originated with the magnetization of a needle using a lodestone. The needle, placed in water, would then always point north. Applications for this were limited, as most land-goers know which direction is north, but on the sea or ocean, the lack of landmarks makes finding a direction more complicated. Sailors would have to use marks in the sky rather than marks on land to find their direction, but the stars cannot always be seen in cloudy weather, especially during long stretches during the winter.

By the time the compass evolved to the point where it could be of some use to sailors, it had evolved into an interesting-looking technology. It needed to be durable, put into a protective casing, and there seemed to be no point having it take up too much space, as a small magnetized needle would point north just as well as a big one would. The compass that Columbus used to sail across the Atlantic in 1492 therefore looked a lot like a pocket watch.

This is of interest because, while it is well known that the mechanical clock led to a "clockwork universe" analogy, we should ask why there was no "universe-as-a-compass" analogy. The compass works with the regularity of gravity, but its source of energy simply seems more mysterious. One has to grind energy into a mechanical clock, watch it dissipate, and then grind more into it.

Meanwhile, cannonballs proved to be useful for blasting through the walls of castles. Cannons, while psychologically terrifying, are not terrifically useful for killing large numbers of people, so the military popularity of these new devices was due to their ability to end the need for medieval sieges. Sieges were nasty bits of business for both the besieged and the besiegers where the objective was either to impose privation and starvation or to survive it.

The cannon created a rough need for physics and engineering that was shaped through practice. No cannoneer sketched a Cartesian graph in the dirt, for the idea did not exist yet. Instead, they realized that by narrowing the cannon, they could shoot the ball more accurately; that the ratio of gunpowder to the weight of the cannonball could be made uniform; and that a 45-degree angle would always send the ball the farthest distance. All of these facts would, eventually, synthesize in the mind of Isaac Newton.

As the cannons blasted, another new sound could be heard in the medieval hamlets: a loud and regular clanging. Great mechanical monstrosities grew from the soil of villages and usually right in the middle of them. The mechanical clock, of course, has ancestors in the form of the sundial and the Middle Eastern water clock. Charlemagne supposedly received a water clock from the Abbasid caliph Al-Rashid (763–809) as a gift, and the notion of the clock quickly spread to the monasteries, where monks used them to keep track of their prayer times.

Alfred Crosby, in his classic book *The Measure of Reality: Quantification and Western Society, 1250–1600* (1997), relates a wonderful diary entry from a monk where, when a fire broke out, some of the friars ran to the river and others to the clock in search of flame-quenching water.

Charlemagne's kingdom featured something that Al-Rashid's did not: winter. Arabic clocks worked by dripping water from a hole in an overhead bucket, and the water collected in another bucket where a floating ball with a pointer on it gradually rose to point at a new number. This is not incredibly efficient, but it is the principle by which modern atomic clocks work, only modern clocks drip radiation rather than water. Cold winters caused the clocks to freeze, so the monks, who were in the process of rediscovering Hellenic forms of engineering from their manuscript collection efforts, began to apply the concept of gears to timekeeping.

While the Chinese had developed mechanical clocks in the eighth century, the technology and know-how remained isolated in the East and seems not to have synthesized into a larger intellectual framework. Monks, rediscovering gear mechanics and used to seeing watermills, eventually captured energy from the environment using gears. The hands of a clock could now turn around a circle just the way that the compass eventually would, and the steady movement of the clock hands created a reference point for the chaotic movement of nature.

The clock's inventors merely took the 12-inch ruler, in use since Roman times, and bent it into a circle. Two passes around the clock face were a close enough approximation to an astronomical day. Why the mechanical clock proliferated in Europe rather than in China or elsewhere is a question without a ready answer. It could be that the European hamlet dwellers were more impressed with the clock towers that suddenly sprang up in their village centers. An impressive clock could draw barefoot rubes from miles away. The clocks got more elaborate, and the gawkers apparently needed more stimulation than just the moving clock hands, so the same principles of energy and motion that moved the clock hands were used to make little robotic versions of Jesus, the disciples, and the devil move.

All of this culminated in the Strasbourg clock, built in Notre-Dame Cathedral in the 17th century. A marvel of early modern special effects, the clock features a cock crowing, Jesus giving blessings, and an astrolabe keeping track of the stars. Stained-glass windows, originally designed to catch the rays of the sun so that the biblical stories could be told in colored light, adorn the background of the clock. The overall effect is to create a feeling of awe in parishioners, something that Christian believers probably associated with religious faith in the same way that Sufi Muslim spinners probably associated dizziness with divinity.

Probably no other technology ever burrowed itself into the collective mind of humanity more than the clock. Mechanized time ripped the sleep-and-work cycle away from its natural rhythms, and pay by the hour contributed to the overwhelming "time is money" metaphor, which, as Lackoff and Johnson have written, defines Western thinking. This is revealed through the fact that both time and money are "saved, wasted, and spent," in the popular vernacular. Clocks created the modern economy but also forced on humanity a perpetual anxiety by empowering hierarchs/bureaucrats to parse out and control certain hours of the day. Both watches and handcuffs go around the wrist.

Scientifically, the clock created a confusion of questions that would take several centuries to sort out. If a ruler measures distance, what does a clock measure? The answer to that is that the clock measures movement, but a full proof of why that is true and the implications of that understanding must wait for the next section of this book.

The traditional narrative has it that the clock created a mechanical understanding of time that divided up movement into quanta and then reified it as an absolute. The thinking became that a clock ticking or a world spinning around the sun could be taken as constant reference points for measuring movement at any other point in the universe, even when the other points are mathematically hypothetical and placed in the distant past.

H

Numbers and the Black Plague

POPE SYLVESTER II (946–1003) ENTERED THE OFFICE OF THE PAPACY IN 999 and tried to bring the Arabic numerals to the Christian West. An intellectual with a fondness for Islamic culture, Sylvester II was ill-suited for the position at a time when the masses judged popes and kings by their military successes and bureaucratic power. His interest in Indian/ Arabic mathematics was theoretical; he enjoyed the intellectual exercise that algebra provided. Yet until Eastern mathematics could demonstrate some application for making money or war, they would lay dormant.

The Italian mathematician Fibonacci (1170–1250) lived in North Africa for a while, where he mastered the use of Arabic numerals, and then wrote a textbook about how to use them that was popularized in 13th-century Italy. Simply put, Roman numerals offered only whole numbers, while Arabic numerals offered the possibility of a decimal system. The more nuanced decimal system allowed for more gradations when calculating interest rates on loans and hence more money. A lender who charged 1.4 percent interest would make more money than a lender who rounded down to 1. If a lender rounded up to 2, then he would be outcompeted by the lower interest rates offered by his competitor.

Via economics, Arabic numerals settled into the intellectual culture of the West. Contemplation of the number 0 and the use of decimal points encouraged mathematicians to think in scales of time that had not previously been considered. Mathematics stretched human thought even as concepts of ratio and proportion, coupled with the rediscovery of the mathematics of Archimedes, made it possible to represent a round surface with a flat surface (cartography) and to contemplate God's creation (as they would have considered it) with a greater level of understanding.

Mathematics might have been put to a conservative cause, that is, brought into a Platonic Christian theology predisposed to accept it, had it not been for the Black Plague. The plague, whatever it was, germinated in the East and spread across Pax Mongolica into Western Europe. The plague was the anarchist concept of creative destruction made manifest; by tearing down the Mongol Empire, it allowed for an East Asian, South Asian, Islamic, and Russian political restructuring. In Western Europe, where the Mongols never conquered, the plague sickened feudal society but did not kill it.

The peasantry, crammed together in village hovels and lacking a knowledge of hygienic practices, died in greater proportion than the nobility. It was not that the nobility was much cleaner than the peasants and not that their royal blood provided some sort of advanced prophylactic against disease but that a greater level of isolation in the manor castles probably made quarantining easier. Even in an era before the means of disease transmission was understood, staying away from anyone with a goiter on his neck and blood in his sputa probably seemed like a good idea.

When the plague itself subsided, a thinned labor force demanded more rights. Economics would predict this. However, something more significant seems to have occurred in that the aristocratic and ecclesiastical powers failed to stop the plague. Frustration with the catholic church found focus in the nascent Protestant movements of Jan Hus and Jean Wyckliff but would not develop a foundation-cracking force until the printing press made the words of Martin Luther available to the masses. By that time, the catholic church had faced a further humiliation: it had not predicted (indeed, no all-knowing holy book did) that two continents full of people could be found at a distance of just a few weeks of sailing.

The catholic church could not stop the plague, it could not predict the Americas, and it had not created the Arabic numerals that suddenly animated economic and intellectual life in the markets and secular universities of Christendom.

I

Descartes, Musical Theory, and the X-Y Graph

SOMETHING MUST BE SAID ABOUT THE TIME IN WHICH RENÉ DEScartes (1596–1650) lived because a correlation seems to exist between political/social upheaval and creative output. Historical causations can be speculated on, not proven, but world history as a discipline allows for speculation regarding causation. Consider these points:

1. Plato wrote in the aftermath of the Peloponnesian War.

2. The Protestant Reformation occurred in the aftermath of the discovery of the Americas, at a time when Native American populations suffered the worst demographic disaster in history.

3. Both Descartes and Thomas Hobbes lived through the Thirty Years' War (1618–1648)—probably the deadliest conflict in human history—and the English Civil Wars.

4. Newton's *Principia Mathematica* was published in 1687, just one year before the Glorious Revolution of 1688 and during a time of English internal strife.

5. Lavoisier's seminal experiments in chemistry occurred during the French Revolution, and Mary Wollstonecraft wrote her *Vindication of the Rights of Woman* during the same time period.

6. Quantum physics, as a theory, was developed in the immediate aftermath of World War I. World War II, more directly, generated a practical application of nuclear science.

Not every period of upheaval leads to an intellectual revolution; it might be argued that a conflict must shake apart the hierarchy or inculcate in the people a distrust of authority for it to create the right conditions for genius. The Crusades, for example, were directed from the top down, and at some level, the papacy's aims were, at least temporarily, achieved. Success strengthens hierarchical controls, whereas disruption creates distrust. A distrust of authority would seemingly be necessary for any new idea to be created and to break through.

Descartes served as a mercenary in the Protestant Dutch military. Mercenaries fought for money, not ideology, and as such were able to exercise more autonomy than a soldier in a nationalist military. Did his time as a mercenary encourage Descartes to challenge or show indifference to authority? Mercenaries don't follow orders mindlessly; they are there to perform a service and survive until payday.

An overlooked component of the Cartesian biography is his connection to music. Pythagoras, unsurprisingly, probably invented the musical scale around 600 BC. That scale, which features notes on a lined graph, allowed for musicians to take sounds (notes) from nature and plot them out in whole-number ratios that, to the human ear, sounded pleasing. When one considers that a chaos of sounds can be made with any musical instrument or combination thereof, this begets a complicated question: "Why do some of these notes in combination sound pleasing, while other notes in combination sound hideous?" There are many more ways to play a violin or piano badly than there are ways to play them well.

One might consider, then, a musical scale to be an algorithm or even an early type of coding. One writes in the commands, based on previous experiences about what audiences like, and those commands create a sound. Music is ubiquitous in the modern world, but in the time of Descartes (or, for that matter, in the time of Nietzsche, who writes hyperbolically about his love of music), it was a rare occasion when someone could hear good music, and a powerful performance could leave musically starved audiences (probably possessed of lengthier attention spans) in a state of rapture for days.

The first book that Descartes wrote, although it was not published until after his death, was the *Compendium of Music* (1618). On page 2 of this short book, one can see his drawings of the musical scale (available at https://archive.org/details/musicaecompendiu0000desc/page/2/mode/2up). The scale, with its points on a graph, looks similar to the Cartesian (X-Y) graph that Descartes would create less than two decades later. The Cartesian graph is fundamental to physics and integral to calculus and forms the central function of algebra. If a musical scale pulls information from sound, turns that into a coding function, and then places the sound on a graph, then a Cartesian plane merely does that with all forms of information. Data act as functions such that "when X is 2, then Y is 4" can be encoded as a function, and this creates an algorithm for understanding the behavior of complex systems.

To use a practical example, given that gravity as a force is calculated at -9.8 m/s^2, one could calculate that 1 pound of gunpowder would shoot a 10-pound ball for, say, 50 yards. If the Y is the amount of gunpowder and the X is the cannonball, then $(1, 50)$ are the coordinate points on the plane. A proportional ratio would then allow a physicist to determine that $(2, 100)$ would be the next points on the plane such that doubling X would also double Y. It works this way in theory. In reality, doubling the gunpowder might shatter the cannon barrel, or gunpowder might flash at a slower rate given the increased density and longer burn. Cartesian graphs don't tell the cannon what to do.

J

Galileo and Bacon

GALILEO (1564–1642) GENERATED THE SCIENTIFIC REVOLUTION, AND Francis Bacon (1561–1626) encoded the scientific method into an educational philosophy. The synthesis of experimentation and education created a foundational philosophy that would manifest itself originally in German research universities, thus encoding and perpetuating the Scientific Revolution into Western civilization.

No story in scientific history is better known than Galileo's. An egocentric professor who, using a pirate-like spyglass as his prototype, learned lens grinding and developed the first genuine telescope, Galileo made his most significant celestial observations in 1609 and 1610. By looking through the telescope first at Earth's moon and then the moons of Jupiter, he not only made discoveries (deep-time celestial collisions pockmarked the moon, and Jupiter has moons that can't be seen by the human eye alone) but also sketched those discoveries in his notebooks.

Galileo published his first book, *Siderius Nuncius* (Starry Messenger), in March 1613. Shortly thereafter, he began to argue in favor of a heliocentric "solar" system. In order to make such an argument compatible with Christian dogma, a dogma based loosely on a geocentric structure superimposed on a Great Chain of Being (Earth being second to the bottom, just above hell, the latter conceit of which is surely drawn from ancient observations of belowground volcanic activity), Galileo had to argue in favor of a more liberal reading of the scriptures, treating the stories in the Bible as allegorical.

It is hard to reconcile the clash that came between Galileo and the Catholic Church with the previous practice of Catholic intellectual life. Since the fall of Rome, the Church had been the great preserver of Western intellectual life. The scribes and theologians of the Christian monasteries welcomed ideas from the Islamic world (indeed, going so far as to choose Sylvester II as pope) and enthusiastically incorporated Aristotle's reasoning into their curricula. That educational system revered secular learning and developed the graduate-degree structure that is still in existence. The Church has never taken a position of biblical literalism, choosing to see the Church's official position on spiritual matters as being divinely rather than biblically supported.

Yet Galileo faced the Inquisition for his public support of heliocentrism. Why? One hypothesis, quite popular, is that Galileo personally insulted the pope in the *Dialogue concerning the Two Chief World Systems* (1632), more than a century after the Reformation, when woodcuts and illustrations equating the pope with the Antichrist were all the rage with certain segments of the Christian population. One would think that Pope Urban VIII, even if perturbed at the betrayal by the astronomer who had once been his friend, could have taken even a mean-spirited joke by that time.

Traditional studies of the time treat the heliocentric model as a threat to the Church's dogma, but this reading of the history misunderstands the nature of bureaucratic structures. A hierarchical entity like the Catholic Church could not tolerate the notion that significant ideas could develop outside the bureaucracy, and this was particularly egregious given that Galileo's ideas did nothing to further any bureaucratic goals.

Since the 18th century, when the term "bureaucracy" evolved (somewhat nebulously) in France by opponents of the ancien régime, the rules and functions of bureaucratic structures have interested sociologists and historians. Max Weber (1864–1920), the most famous of German sociologists, shaped definitions regarding the functions and culture of bureaucracy. The hierarchy in such systems involves decision-makers parceling out duties to underlings. Power is defined by the relationship between the person giving out the duties and the person who does them.

From Weber's foundational thesis, it's possible to view the Catholic Church as just another bureaucratic institution. If bureaucratic structures define themselves by the asymmetry of the power relationship between the person handing out tasks and the person doing the tasks, then rewards and punishments remain comfortably within the purview of the bureaucracy. Underlings are supposed to accept tasks, perform them, and then be judged on that performance by their superiors.

Bureaucracies stifle ingenuity by redirecting any human impulse toward standardized mediocrity. The earliest theologians of the 12th, and 13th, and 14th centuries featured some of Western civilization's most impressive thinkers in the form of Grosseteste, William of Ockham, Abelard, and Aquinas, but when a Scholastic bureaucracy hardened around these ideas, a stasis set in, and the Church educational system produced nothing of intellectual value at all from the 14th century on.

Theology after Ockham developed into the kind of bureaucratic legalese that plagues modern corporations, where intellectual work consists of studying ever-denser forms of regulations that increase either the profit of the company or the power of a bureaucratic wing, with the two goals often working in tandem.

When underlings perform tasks not assigned to them and when those underlings receive recognition for those tasks, especially when the task works against the goals of the bureaucracy, then discipline must be meted out. One could not imagine, for example, that a middle manager in an insurance company who, on his own initiative, published a mathematical theory about how to lower premiums, reduce red tape, and pay for more services would be promoted within the company's structure. Instead, he would likely be either ignored or forced to explain himself to human resources.

Unfortunately, for Galileo, the baroque-era version of human resources was the Inquisition. The Church hierarchy, consisting of the same ageless and faceless bureaucrats who populated it at the time of Luther more than a century before, reacted to Galileo in the same way that they had reacted to Luther: the bureaucracy pressured both to recant. To compare the responses of both men to this pressure is to see what the

116 years between 1517 and 1633 had meant for the influence of the Catholic Church.

When brought before the Diet of Worms in 1524, Luther refused to bow before the assembly of secular and religious big shots who pressured him to redact his message of religious reformation. By that time, Luther had already been excommunicated for three years and had developed a theory about the Church that rewrote the bureaucratic/theocratic rules to favor Luther personally while also upholding the monumentality of the struggle that Luther found himself engaged in.

Traditional Christian belief held that the Church was God's representative on Earth; that the pope stood atop of a hierarchy of cardinals, archbishops, priests, and so on; and that only God stood above the pope. This hierarchy oversaw the Christian mass, and a series of rituals and sacraments ensured an individual that he or she was inside the mass and therefore eventually would pass into heaven. To be excommunicated from the mass meant, theologically, to also be removed from the rolls of those about to enter heaven, and those who died outside of the Church's mass would suffer from hell's eternal torments.

Luther developed a theology that declared the Church to be hell's representative on Earth, and therefore the pope became the representative not of Christ but of Satan, making the pope the Antichrist. To be excommunicated from the Christian mass therefore was to be saved and not damned as long as the individual believer possessed faith alone. When brought before the Diet of Worms in 1521, Luther saw himself battling against an institution possessed of demonic strength, and he believed that the souls of faithful Christians were at risk. He saw his adversaries as being spiritually formidable, and his response was equivalent to that belief. The Church held spiritual authority (even if demonic), and only a person of significant spiritual strength could fight the Church on a celestial scale.

While the Church bureaucracy was unlikely to be pleased by such a doctrine, at least Luther assigned them a substantial role in his morality play. The pope asserted that he was Christ's earthly representative, and Luther argued that the pope was the Antichrist. Religious believers could choose this new interpretation without threatening the background scenery of the play; God, the angels, and eternity still hung in the back-drop, and souls were still eternal.

Luther damned the Church with his refusal to recant, but Galileo damaged it more significantly with his apology. The Catholic Church is just an earthly bureaucracy, in no way representative of a spiritual world that exists only in concept. If it wanted to bully an old man into recanting an irrefutable astronomical observation, that moons orbit Jupiter, and that those moons can be seen only with sight-enhancing technology, that would not in any way change the fact that Earth orbits the sun.

By recanting, Galileo treated the Church as an absurd bureaucratic institution, one that contributed not at all to his great discoveries and that had no power to put the sun in orbit around Earth. Had Galileo taken a stand, per Luther, for scientific truth, he would have broken through torture on a planet that spun around the sun. If he recanted, he would be apologizing for falsehood on a planet that spun around the sun. In Galileo's scientific play, the pope is not even a character, and physics takes the place of the angels.

The historian could imagine a situation where the immediate implications of Galileo's findings would have been ignored by European thinkers for decades or centuries. The question of celestial rotation was likely seen as an internal intellectual dispute with few genuine implications for the getting on of society. Crops needed to be harvested, and babies fed no matter what rotated around what.

Yet Francis Bacon (1561–1626) swiftly recognized the implications that Galileo's 1610 discovery should have for understanding the history of philosophy and for the restructuring of education. By 1620, Bacon published *Novum Organum* (New Organ, or New Method) and argued that Aristotle's philosophies (the title is a play on Aristotle's classic *Organon*) should no longer feature at the core of Western education's educational structure. The purpose of the intellectual enterprise, of which education is a part, should be to create new knowledge (as Galileo had through experimentation) rather than just to study old knowledge.

Bacon had been a contemporary of William Shakespeare, and the two share a dramatically important feature. Both employed their talents at and for the pleasure of a monarch. Shakespeare never worked as a government official, but Queen Elizabeth often featured as a very important person in his audience. Likewise, Bacon's talents belonged to King James I.

A century of Protestantism in England must have created a new relationship between the ruler and ruled, or maybe it was the way in which "Good Queen Bess" had made her queenly appearance before the English commoners in 1588, commanding the winds to blow away the Spanish Armada, but the English church and crown seemed open to rather than threatened by works of genius. Elizabeth adored Shakespeare, and Bacon dedicated *Novum Organum* to his king and personally presented the sovereign with a copy.

This is a subject that has so far gone unconsidered: Galileo thought within the Catholic intellectual universe, and the bureaucracy abused and suppressed him. Bacon, who operated within the English government and therefore under the direct authority of the English king, presented a spell book, sparkling with new ideas and ingenious criticisms, to his king and suffered no reprisals. King James, author of tracts condemning both tobacco and witches, was as close to an intellectual as any monarch is ever likely to be. He did not fear the genius in his employ. When Bacon fell from political favor, it was for embezzling funds, not embracing new ideas.

K

Isaac Newton

THE 46 YEARS BETWEEN THE PUBLICATION OF *NOVUM ORGANUM* AND Newton's creation of the calculus were not without scientific advancement, but it must have been hard for anyone to think about math through the rattle of cannons and the screaming of victims. The Thirty Years' War, fought between 1618 and 1648 in Germany, likely killed more people as a percentage of the population than any other of history's conflicts. At the same time, the English Civil Wars (1642–1651) pitted Puritan parliamentarians against an Anglican king, Charles I.

The two wars together brought about an end to the concept of any European religious empire. In 1635, the Catholic French entered the Thirty Years' War as antagonists to the equally Catholic Habsburgs, thus ending the conceit that the soldiers involved fought for religious glory. Likewise, the English Parliament increasingly justified its Civil War against Charles by invoking the Magna Carta of 1215 and eventually convicted Charles of treason, not heresy.

Newton's age was one that was weary of religious strife and leery of religious authority. Newton himself, an amateur theologian and all-around mystic, disbelieved in the Holy Trinity. This might seem a minor sin, but he privately harbored a thought that would have upset both the Anglican and Catholic God in the 17th century.

Biographical treatments of Newton tend to divide his life into two parts. First, there is his lonely childhood, his late arrival to Cambridge followed by an early and temporary exit during the plague year of 1666, his development of the calculus, a theory of "optiks," and the inverse square law. Having engaged in such an impressive initial burst of mathematical creativity, he put his thoughts away and spent nearly two decades as an obscure lecturer in mathematics.

Then, in 1684, Edmond Halley (the closest thing Newton had to a friend), fresh from a coffeehouse discussion about the shape of the planetary orbits around the sun, paid a visit to Newton's lonely rooms. While there, plopped down among Newton's books and papers, Halley asked if Newton knew what the shape of the planetary orbits were. Newton replied that the orbits were elliptical and that he had worked the problem out long ago but would have to re-create the theories and equations because he could not locate his original work.

Although an abstract was mailed to Halley not long afterward, by 1687, Newton's *Principia Mathematica* saw publication. In a fact too remarkable to be coincidental, a year later, the English Parliament would invite help from the Protestant sovereigns of the Netherlands (William and Mary, who happened to be the son-in-law and daughter, respectively, of James II) and seized power from the sovereign in a Glorious Revolution. Western civilization, lacking a central pillar of authority since Martin Luther destroyed Christendom in 1517, now had two new pillars of authority in the forms of science and democracy.

Newton certainly possessed a singular genius, but the tumult of his era must have broken apart authority in such a way that he could conceive of himself, rather than the accepted wisdom of all of science, as correct. In the next century, Adam Smith's *Wealth of Nations* and the first volume of Gibbon's *The History of the Decline and Fall of the Roman Empire* would be published in 1776. Then the law of conservation would be founded by a Parisian, Antoine Lavoisier, in the inaugural year of the French Revolution. Mary Wollstonecraft, who joins Claudius as Western civilization's only two genuine geniuses between Newton and Einstein, created feminist theory.

The collapse of authority in the West, when contrasted to the more stable and stagnant civilizations of the Islamic world and Asia, must have created conditions amenable to creative thought. The notion might have the effect of permanently condemning hierarchical structures and again gives some insight about why early "natural philosophy" came across as so dangerous to authority.

Newton's relevant ideas, created in the initial 1666 burst of creativity and then in a second burst of genius 20 years later, can be labeled and explained succinctly:

The First Law of Motion

Every body perseveres in its state of rest, or of uniform motion, in a right line unless it is compelled to change that state by forces impressed thereon.

The Second Law of Motion

The alteration of motion is ever proportional to the motive force impressed and is made in the direction of the right line in which that force is impressed.

The Third Law of Motion

To every action is always an opposed and equal reaction, or the mutual actions of two bodies on each other are always equal and directed to contrary parts.

The Conservation of Momentum

If particle A pushes on particle B, the two forces are equal in magnitude, which means that P1 + P2 = a conservation of momentum, or a constant.

Calculus

Newton likely created calculus just before the German philosopher Gottfried von Leibniz did so independently at just about the same time. Mathematicians since the ancient Greeks recognized (they expressed the recognition through paradoxes) that taking an average speed over a distance was insufficient for solving certain problems regarding motion.

An apple dropping 9 feet in 9 seconds falls, on average, 1 foot per second; it will not actually fall at that rate, as the speed of the apple will increase as it moves toward the ground.

Newton developed the concept of a limit to state the idea of the "approach." If the ground represents "9" (being the final stopping place of a 9-foot drop), then an apple rising and falling on a quadratic curve would be x^2, with a limit of 3, meaning that as the apple starts to approach a speed of 3 feet per second, it is about to hit the ground.

The concept of a limit in calculus, like the notion that one must draw smaller and smaller rectangles to derive the area under the curvature, is based on the conceit of infinity. Infinity as a concept creates paradoxes, but mathematicians tend to use it as a shorthand for stating, "You can see that this principle will hold true even for a really big number."

The Inverse Square Law

A physical entity is inversely proportional to the square of the distance. Newton conceived of this as a mathematical inference derived from studying conic sections. The theorem of the law is usually explained geometrically through the use of ellipses, arcs, circles, and/or hyperbolas. Practically speaking, Newton used the inverse square law to describe how the force of gravity could pull something downward and why the intensity of that force increases as an object reaches a limit (the ground or the moment just before the object hits the ground).

Earth's gravity pulls down with a force of 9.8 g's. However, if, by propulsion, an object can rise off Earth, then the force of gravity gradually relents the higher that the object goes. This concept works in forward or reverse. An object thrown out of an aircraft two miles in the air will not have any propulsion to counter the force of gravity pulling it downward, so while gravity might not be as powerful, say, two miles upward, it is the only force working on the object. As it falls, gravity's force increases until the full impact of 9.8 g's can be felt on the ground.

Conversely, rockets start on the ground feeling the full force of Earth's gravity, but the propulsion of the fuel sends the rocket upward, and less and less fuel is needed the higher the rocket goes. If the rocket gets far enough away from Earth, then it avoids the gravitational pull and can move toward other celestial objects.

Newton's primary concern was in explaining how gravitation counters the force of spin, or centrifugal force (there is some argument as to whether centrifugal force is a force at all or just an object trying to move in a straight line, becoming discombobulated when the straight line keeps moving).

In short, people wanted to know why they didn't fly off the planet if Earth was spinning on its axis. The reason for this is because Earth's size creates much more gravitation force than the spin creates centrifugal force. At the equator, where the centrifugal force is the strongest, it still only counteracts about 1 percent of Earth's gravitational force

Earth "weighs" 13 thousand, 170 billion trillion pounds, and the spin of Earth at its maximum represents 1 percent of that force. So if we kept Earth spinning at the same speed but reduced its weight by 99 percent, then a bunch of stuff at the equator would start to fly off. Likewise, if we keep the weight of Earth the same but increased the rate of spin by 99 percent, then that same stuff would also start to fly off.

The mathematics for this is not complicated, as it could be calculated with simple ratios, but the inverse square law was conceived in tandem with calculus, and the rates of fall and limits were written in mathematical language involving radii and in the precise language of trigonometry. Like all concepts in mathematics and physics, the inverse square law can be conceptualized easily but understood precisely only through the use of calculations.

Functions

Cartesian graphs with bullet points originated as a graphical representation of cannonball trajectories. If a historian of science wanted to be as ungenerous as possible to Isaac Newton, it could be said that he provided mathematical precision to the concept of the elliptical arc, which was already well understood by cannoneers.

By the 15th century, gunpowder technology had become embedded into European warfare. Some archaeologists even believe that there is evidence of "firearms" (what might be better called "hand cannons") being used in 1415 at the Battle of Agincourt.

In some ways, the use of gunpowder on the battlefield is a bit of a puzzle, as the harquebuses and muskets used in the early modern period hardly seem effective at killing people. Even in the early 17th century, when John Smith, while acting as a military protector for the Jamestown colony in North America, found himself a captive of the Powhatan tribe, he broke his rifle rather than engage in a shooting contest against an Indian with a bow and arrow. Smith knew that his rifle would lose.

Gunpowder weapons had to pass through a phase of evolution where, for a time, they were not effective for killing people. This would be akin to the period where dinosaur feathers had to be used for attracting mates, warmth, or displays of aggression before they could eventually evolve to be useful for flight. During this phase, gunpowder weapons initially scared horses, as the Mongols supposedly stuffed snippets of bamboo with gunpowder and used them as firecrackers to scare the enemy horses.

Rudimentary cannons, or bombasts, had barrels that were too wide, so the blast wasted energy that otherwise would have fired behind the cannonball, thus rendering the bombast ineffective as an actual weapon. However, the sound itself and the rumbling of the ground after the explosion was apparently enough to cause entire villages to surrender during the last stages of the Reconquista on the Iberian Peninsula.

As was stated earlier, cannoneers of the late Middle Ages "played" with the angle of the cannon until they came to an equilibrium. Game theorists describe the concept of "playing" to an equilibrium through a variety of simulations. One simulation involves asking each individual in a room of, say, 20 people to pick a number between 1 and 100 at random. Anyone who gets the closest to one-third of the average of the number will win a prize. At first, the bewildered players probably think the game is based on random chance, and some will pick high numbers and others low numbers based only on random distribution. (Hardly anyone can actually apply mathematics, so the fact that the highest possible number is "33" [that is, if all 20 people picked 100, then one-third of the average would be 33] will likely go unnoticed.)

If the game is played enough, even players unfamiliar with mathematics will discover that one-third of the average of the numbers picked tends to be a lower number. Players will naturally gravitate toward lower and lower numbers until everyone realizes that one-third of the average of 0 is always 0. If all 20 players just write down a "0," then they will all win every time. Anyone who plays above a "0" will be the only loser.

The game naturally developed to a Nash equilibrium, and then a mathematical model was superimposed on what naturally happened. In almost all cases where there is some kind of competition, there is an early learning process where players work through strategies that are successful, then eventually balance out their strategies and come to a kind of competitive stasis where outliers (such as incredibly talented athletes or those who are willing to use performance-enhancing drugs) determine victory.

The reason that the mathematics of game theory became necessary was that humanity cannot afford a "learning phase" when thermonuclear weapons are in use. That's a game that will be played only once, so a mathematical theory that brings the "players" to a state of equilibrium before the war games start becomes essential.

All of this is to say that the quadratic curve, the elliptical movement of spheres, had already been worked out by illiterate medieval cannoneers, but Isaac Newton explained the concept on a Cartesian graph. Then he extrapolated the numbers out to show that the same concepts of gravitation that work on small spheres (cannonballs) also work on big spheres (planets).

The mathematics for this is not complicated and mostly involves developing simple ratios. If one works out, for example, that a 10-pound cannonball fired from the ground at a 45-degree angle will travel 100 yards, then one can take that experimental finding, apply the inverse square law, and determine how far the cannonball would fire if it were 20 pounds.

Also, given that force = (mass)(acceleration), one could determine that a 10-pound cannonball traveling with an acceleration of 10 miles per hour would exert 100 pounds of force, just as a 20-pound cannonball moving with an acceleration of 5 miles per hour would. Double the weight and halve the speed, and the force is the same. Halve the speed and double the weight, and the force is the same.

Simply put, the arc of a cannonball is a partial orbit. Go up high enough so that the power of gravity is reduced, and the cannonball will go around Earth in the shape of an ellipse and become a satellite.

Over time, the input value on a Cartesian graph, used by Newton to calculate the trajectory of spheres and their relation to gravity, turned out to be useful for plotting abstractions such as "consumer confidence." The "bullet" points on any graph that one encounters originated as cannonballs just as surely as the (false, as we shall see) notion of a "big bang" is also derived from the superimposition of cannonball physics onto the origins of the universe.

That Newton was able to live the second half of his long life enjoying the wealth and fame derived from his creation of the calculus and laws of physics is integral to understand his long-lasting impact. He created conditions by which subsequent thinkers, aspiring to be considered geniuses, also tried to explain complex phenomena using a small number of explanatory premises.

Adam Smith described economics in terms of free supply and demand, and Thomas Malthus applied mathematical principles (slow addition for agricultural land and quick multiplication for birth rates) to population growth and decline. Darwin built on that to describe the internal workings of that population growth and decline through his concepts of natural and sexual selection. Georg Hegel would describe history through the thesis, antithesis, and synthesis, and then Karl Marx would (over)simplify to the conceits of two economic classes in eternal struggle. Sigmund Freud chose the id, ego, and superego as explanatory factors for the concept of human behavior. All of these thinkers aspired to Newton's genius, but only one achieved it.

Only Mary Wollstonecraft would ever achieve a Newtonian level of thought. Civilization, wrote the lady, is the extrapolation of the natural male physical strength over women, an insight that, once seen, forever leaves her readers in a new intellectual state. After one reads Wollstonecraft, one gets a new set of eyes, and so too was it with Newton.

L

Binary Code and the
Law of Large Numbers

GOTTFRIED VON LEIBNIZ (1646–1716) DEVELOPED BINARY CODE IN 1689, not incidentally (the year that the English government created the first "bill of rights"). In a short paper titled "Explication de l'Arithmétique Binaire," Leibniz spelled out the concept of the binary code, a system of simple 1s and 0s. Because of the feud over who invented the calculus first, both Newton and Leibniz have developed slight reputations for pettiness, but in the "Arithmetic of the Binary," Leibniz shows no evidence of this personality characteristic. He goes out of his way to credit an ancient Chinese philosopher for intuiting the concept.

Leibniz recommends binary code because, unlike the times tables, its usage required almost no memorization. Binary code works by writing traditional numbers and then turning those numbers either "on" or "off" with a 1 or 0 in a box by the numbers. For example, if the real numbers 2, 4, 6, and 8 all had a "1" in the box by them, then this would represent the number 20. If the "2" had a "0" in the box and the other numbers had a "one," then this would represent the number "18."

Leibniz did not see this, but binary code actually reduces the concept of infinity and shrinks the memory capacity of information-carrying particles in a mathematical sense. However, its practical use is enormous given the need for high levels of memory capacity. The traditional narrative has it that the development of machine computing in the mid-20th century caused computer scientists and mathematicians to reach back in time and synthesize binary code with the memory capacity of modern computation. This is correct, of course, and of great importance for the modern world.

However, for the thesis in use here, the structure of binary code is more significant. Binary code separates out the 1s and 0s and corresponds them to other numbers, which could be considered a state of being, such as temperature. This caused a new evolutionary branch in mathematics that would eventually develop into matrix algebra. Matrix algebra, like binary code, assigns numbers to states and therefore, unlike binary code, is not commutative. Binary code can be run forward and backward and give the same mathematical result, but matrix algebra cannot. The development of noncommutative mathematics will be fundamental to the development of quantum physics and, eventually, the long-overdue death of Plato.

In 1713, a Swiss mathematician named Jakob Bernoulli (1654–1705) created the law of large numbers and therefore founded the mathematical field of statistics. The law of large numbers relates to the disjunction that sometimes occurs between what a mathematical model might predict and what actually occurs when events play out.

If a statistical model is a thesis, then randomness acts as the antithesis. A statistician might work out that the odds that a "person-on-the-street" interview with a randomly selected American would reveal that Americans cannot find Mexico on a map 25 percent of the time. It is entirely possible, however, that the actual results of any given person-on-the-street interview would find that only 15 percent of Americans can identify Mexico in one neighborhood and that maybe 100 percent could in another neighborhood.

All localized anomalies get ironed out, at least in theory, as more and more neighborhoods are sampled, with the numbers eventually converging on what the statistical model originally indicated. This does not always work and is sometimes labeled as a frequentist fallacy, but it works enough to make money from the concept. The law of large numbers became the basis for the insurance industry, with insurance companies collecting premiums from vast numbers of people. Each of the individuals who takes out insurance is hoping that nothing catastrophic will happen to him or her, but each needs insurance just in case.

Meanwhile, the insurance companies capitalize on that need under the mathematical assurance that the law of large numbers will keep the number of cancer patients or destructive floods at a minimum.

Money is to be made in between the numbers, as a company constantly needs to adjust premiums as the situation, particularly the climate, changes.

M

The Great Era of Mathematics

Despite Aristotle's disruptive and productive influence on medieval European thought, Plato rules the Scientific Revolution until the early 20th century. The dominance of Platonic thought is revealed by Galileo's famous declaration:

> [The universe] cannot be read until we have learnt the language and become familiar with the characters in which it is written. It is written in mathematical language, and the letters are triangles, circles and other geometrical figures, without which means it is humanly impossible to comprehend a single word.

Plato's philosophy of mathematics, derived primarily from geometry, had it that perfect shapes existed in a Realm of Forms. Such a philosophy could support a mathematical connection of any kind with science. It is, for example, possible to imagine a melding of Platonic philosophy with the Chinese and Japanese "brush form" of numerals because the conceit of the numbers is the same: these are markings on the page that stand for a certain number of objects, such as animals, houses, or debt, that exist in the world off the page.

Any historical numerical set could channel Plato's philosophy, but some made for more effective conduits than others. Roman numerals scrambled or blocked the Platonic message in Western civilization for centuries. The arrival of Hindu-Arabic numerals, however, provided a perfect vessel for Platonic philosophy to travel through.

Before the numbers 0 to 9 went by the name of Arabic numerals, their Indian creators referred to them as Gobar numerals, meaning "dust numerals" because in order to work with the numbers, mathematicians simply inserted a forefinger into the dirt and drew the numbers. "Dust numerals" is a superior phrase to Arabic numerals because the term captures the Platonic conceit as one might think of the Realm of Forms being made of a dusty substance, hazy in the mind.

Dust numerals proved to the best conduit for Platonic philosophy as mathematics quickly came to be seen to have special philosophical properties. It was as if—and to paraphrase Galileo—the natural philosophers discovered the exact abstract concept that described the universe.

Mathematicians who wrote in dust numerals could ascertain the circumference of Earth and the elliptical shape of the planets and not only understand the force of gravity on objects on the ground but also calculate accurately the intensity of gravity as it would act on those same objects should they move into the atmosphere or beyond.

When the calculated predictions fit, almost exactly, with the observable phenomena of the universe, one could interpret the miniscule difference between mathematical model and actual evidential results in one of two ways: either the evidential (real) world is wrong and fails to match up to the Platonic world, or the evidential (real) world is right, and the Platonic world fails to measure directly up to it.

A philosopher, natural or not, could probably have made either case by the middle of the 18th century. However, as events would unfold after this, the development of probability theory would indicate an increasingly large gap between the evidential and Platonic worlds. This gap would eventually force mathematicians to explain themselves. If mathematics was not the language of the universe, then it must be a logical system, and if it is a logical system, then what is its foundation? If it has no foundations, then the surface-level perfection of mathematics can be understood only as a series of tautologies.

Such a logical crisis will eventually create the Vienna School and eventually an uncertainty theorem (1931) by Kurt Gödel, who was the first mathematical philosopher to get the dust out of his eyes. Gödel's incompleteness theorem must later be discussed with greater context;

however, his unease over the foundations of mathematics is predated by Heisenberg's uncertainty principle (1927), which itself was preceded by the development of matrix algebra.

Heisenberg needed to employ matrix algebra because the algebra available to Western mathematicians, brought to Europe by Girolama Cardano who wrote *Ars Magna*, which translates to the *Great Book* (the reprint publisher, recognizing that a more descriptive title might be necessary for book browsers, called it *The Rules of Algebra*) was insufficient. *Ars Magna* contains an error as significant for the history of mathematics as the Ptolemaic error (passed on through Aristotle) regarding a geocentric planetary system was for cosmology. That error is known as commutation.

The commutative property existed long before Cardano, and no mathematician can claim to be its originator. Probably, commutation seemed to derive from "common sense" in the same way that a moving sun and a stable Earth did. If Earth appears stable and the sun appears to move, then why assume otherwise? If you have four apples and you rearrange the order of those apples, there will still seem to be four. Codifying such a concept into numerical representations of those apples would seem to be an obvious evolution from the sensory experience to mathematical representation.

Nonetheless, Cardano directly expresses the commutative property in two ways. In chapter VII, "On the Transformation of Equations," he writes,

> When a number and a middle power are equal to a higher power, the equation may be converted to one of the same two powers, with the same coefficients, equal to the number. As, if I say,

$$X^2 = 6x + 16$$

> We can also say

$$X^2 + 6x = 16$$

And conversely. Then, having the solution for the first of these, we subtract from it the coefficient of x, which is 6, and we will have the solution for the second, or, having the second, we add to it 6, the coefficient of x, and thus make the solution of the first. Truly, [however], no general rule can be given for the other powers. (56)

More directly, in chapter XXXI, "On the Great Rule," Cardano writes,

> This is a rule for solving great problems and from it were discovered the rules for alloys of gold and silver. It incites ingenuity and does so through demonstration. It requires an expert and is [best] taught by problems since it is many-faceted. Commutation is the foundation of the rule. (186)

The commutative property's central part in algebraic mathematics was based on the idea that Pythagorean wholes (one apple or one rock) would be unchanging and could be described in a world of Platonic mathematics. The whole apple was an imperfect shade of a perfect apple that existed in another realm, and in that Platonic realm, you move numbers around without altering the effects of subtraction, multiplication, division, or addition. The historical benefit of this was to create a conceit in mathematics that made common sense based on the sensory evidence available during the Scientific Revolution. The error comes from the inherent belief that in the Platonic realm, nothing radiates.

N

Revolutions and the Carnot Cycles

THE SECRET HISTORY OF THE MONGOLS, WRITTEN ANONYMOUSLY AND almost certainly designed to provide the Mongol emperors of the mid-13th century with an easy-to-understand origin story featuring a "great man," relates the story of a young Mongol child named Temujin and his rise to become the Genghis Khan of the Mongolian horse nomads. *The Secret History* contains a story as unsettling, in its way, as the biblical tale of Abraham's near sacrifice of his son Isaac.

Temujin was supposedly the son of a woman and her kidnapper. Mongol society consisted of various clans where status was assigned by various levels of wealth (measured largely in horses) and military capability. Men who existed outside of those societies roamed the countryside in small packs and survived by hunting, stealing, and kidnapping. Although women in Mongol society typically enjoyed rather high status for the era, a fact partially evidenced by the fact that Mongol women never had to endure foot-binding, they were commonly kidnapped and abused as spoils of war.

Temujin's mother, Hoelun, barely got to enjoy her wedding before being kidnapped by a rogue horse warrior, thus becoming a victim of the steppe's lawless bandits. Hoelun was stolen, and her son, Temunin, was the child of her kidnapper. Temujin's future greatness was foretold at his birth as he came screaming from the womb while holding a blood clot in his fist.

A rival clan executed Temujin's father at a time when Temujin himself was only a preadolescent. According to the customs of the steppe, this meant that Temujin's older half brother now became the head of the family and therefore had rights to anything that Temujin acquired through stealing, hunting, or fishing. One day, when Temujin caught a fish, his half brother took it. Unwilling to accept his subordinate status, Temujin cajoled his younger brother into distracting the older half brother, and Temujin then shot his older half brother in the back with an arrow and left the body to rot in the open air.

The incident might be read better as a fable than as a history, the lesson being that Temujin did not intend to fight his enemies in a "fair" way. Why should a smaller and younger boy fight an older and stronger one in hand-to-hand combat? Why not, instead, establish the rules for a sneak-up-and-shoot-you-in-the-back contest and then win before your opponent even knows that there is a game?

The Temujin fable also establishes the conceit that warfare creates conditions where creative thinking, often coupled with amorality, provides direct benefits for the creative thinker. The traditional historical narrative treats warfare as an impetus for technological innovation and progress in military engineering, but warfare and upheaval might also have the effect of disrupting norms of behavior in a way that leads to cunning and creative new patterns of thought.

The 18th century brought revolutions against a mercantilist colonial system (the American Revolution), against a medieval and extractive political structure (the French Revolution), and against the plantation slavery (the Haitian Revolution). For the American Revolution, it is hard to see the Enlightenment as a causal factor, as the American colonists seemed driven to revolt more by economic and territorial ambition than by philosophical rhetoric, but the revolution itself opened up an opportunity for Jefferson and Madison to synthesize Enlightenment philosophy into a workable plan for a democratic republic.

The French Revolution, being in the old world, proved to be vastly more disruptive than the American. French revolutionaries could not throw the seeds of philosophy into new soil; they had to hack through the roots and rocks of a medieval political and religious system while at the same time turning their planting soil into a graveyard for the clergy and nobility.

The French Revolution overthrew traditional concepts of hierarchy, and if the priesthood could be abolished, the nobility beheaded, and the Christian calendar discarded, then why must women accept the same old patriarchy in a whole new world? Wollstonecraft's *A Vindication of the Rights of Woman* (1792) put forth a thesis more radical and intelligent than anything inked by Locke, Hobbes, or Rousseau. The thesis is this: *civilization is an extrapolation of men's physical strength over women.*

It's a thesis shot right into the middle of the back of Enlightenment historiography. Locke, Hobbes, and Rousseau, with their assertions that men came out of the womb either blank, evil, or good, respectively, could only ape the central conceit of Christian theology. Christian theology, be it Protestant or Catholic, held that humans were born bad and could be cleansed of their sins only through either good works (Catholicism) or individual salvation (Protestantism). Wollstonecraft discarded the entire notion of human nature by recognizing that when humans stopped living as "savages" and started settling in cities, a settled lifestyle favored the physical strength of men and cornered women into a specific societal niche.

Wollstonecraft's daughter, also Mary (last name Godwinson), was born in the first year, 1797, of what might be called the Napoleonic era. Her biography remains one of the most famous in the history of literature, and she published in 1818 the gothic classic *Frankenstein, or The Modern Prometheus.* Due to her marriage to a poet, Mary now carried the last name "Shelley."

Mary Shelley's book functions like few other novels; originally written by firelight, it continues to be read in fractals. The novel can be broken down and rebuilt for each generation. A gothic study of human nature in the 19th century becomes a tale about the consequences of genius without boundaries and the power of science in the atomic era.

The father–son companion to the mother–daughter genius of the Marys comes in the form of Lazare (1753–1823) and Sadi (1796–1832) Carnot. The force of their genius altered the course of Western civilization in ways still not fully comprehended, perhaps because their ideas and force of personality cause changes in so many disparate fields.

Lazare was not yet 40 when elected in 1792 to the National Convention in revolutionary France. A member of the notorious Committee for Public Safety, he became the architect of the first formal military draft in world history (the *levée en masse*), which was instituted in 1793. Few concepts altered the history of warfare like the conscription army. Governments now had to train as well as feed and clothe masses of the lower class.

Conscription created a new nationalistic contract between the government and the people. Governments would now find it in the necessary interest of the state to provide basic services to the people (now potential soldiers), and the people suddenly understood themselves to be necessary for upholding the power of the state. This is significant because soldiers could sabotage a government's ability to wage war in a number of ways, perhaps most significantly by just not fighting hard enough.

It is perhaps too often forgotten that Napoleon's considerable miliary genius could flourish only in an environment where troops *wanted* to fight because they considered themselves to be part of a nationalistic experiment worth holding a rifle for. French troops marched forward with *elan*, or "fighting spirit," in a way that soldiers coerced into fighting for a despised nobility never could. Eventually, the great powers of Western Europe would have to also institute conscription armies and found that if they wanted military prowess, then they would have to concern themselves with the welfare of potential draftees. The draft begat the welfare state.

The times favored the cold-blooded, and Lazare Carnot was certainly that. Weary of Robespierre's tactics and wary of the lawyer turned dictator's intentions, Lazare participated in Robespierre's overthrow and execution. Then, when the political fallout got too radioactive, Lazare had to flee the country in 1797. He never lost his belief in revolutionary ideals, however, and returned to France to serve as Napoleon's minister of war. Like Beethoven, Lazare lost faith in the Corsican general after 1805, when Napoleon crowned himself emperor of France in the Notre-Dame Cathedral. Nonetheless, Lazare, like many others, found himself periodically swept back into the service of France by the irresistible personality of his general and emperor.

As is so often the case in philosophical history, the chaos of the era seemed to be a stimulant to Lazare's intellect rather than a hindrance. Indeed, one of the least appreciated advances of the time, if only because its creation coincided with the Reign of Terror, was the establishment of the École Polytechnique in France. Whatever the revolution's excesses, it was effective at tearing away a medieval political, military, and educational establishment that had long protected its own privileges at the expense of progress.

This tearing down of medieval institutions in France took place at an opportune time, for the Germans (Germany being the name of a region until 1871) developed the modern research-based doctorate in the late 18th century and provided the appropriate model for the French. The revolutionary government needed army engineers, and the École Polytechnique immediately attracted, educated, and facilitated a mass of impressive men. This new university acted as a catharsis, as France teemed with thinkers who were interested in the forces driving this new era of steam and iron.

In his introduction to a reprint of Sadi Carnot's *Reflections on the Motive Power of Fire, and on Machines Fitted to Develop That Power* (2005), Mendoza writes of the university and its effect:

> A list of men associated with its early period, before 1830, reads like the index of a book on mathematical physics. Among the first instructors were Lagrange, Fourier, Laplace, and Berthollet, Ampere, Malus, and Dulong; among former students who stayed on as instructors were Cauchy, Arago, Desormes, Coriolis, Poisson, Gay-Lussac, Petit and Lame; other students included Fresnel, Biot, Sadi Carnot, Claperyon, and the physiologist Poiseuille. It was this generation which largely formulated the attitudes and procedures of mathematical physics as we know it today. The terms Lagrangian, Laplace transform, Poisson's equation, Fourier integrals, Cauchy relations, Fresnel coefficients, Coriolis forces conjure up a picture of an era when it seemed that the understanding of nature was merely a question of writing down equations and of inventing methods for solving them.

But there was another aspect of the world of this extraordinary group of scientists. Possibly because of their connection with an army which conquered most of Europe, possibly because they lived in such a time of discovery and transition, almost all of them saw their science as only one part of their lives. While being masters of pure science, they applied their astonishingly varied talents to very different problems. (viii)

Such an environment helps to explain how Lazare Carnot could both invent the concept of the military draft and, in 1803, publish a work on projective geometry, the practical effects of which can be at least partially (or fully by someone who is a projective geometer) seen in the fortification now known as a Carnot wall.

In fact, the Carnot wall might be seen as the practical military manifestation of two of Lazare's converging ideas. In 1784, Carnot published a paper showing how kinetic energy can be lost in certain types of collisions. The Carnot wall depicts a geometric design where the outer wall of a fortification looks like a star. The only solid point where a cannonball could land would be on the thin outer edge of the star. If a projectile lands anywhere else, it will skid and lose kinetic energy in the same way that a modern skateboarder will skid rather than slam if he falls off his board in the middle of a ramp.

The Carnot wall is a triumph of thought if one considers that by 1803, gunpowder cannons had been knocking down castle walls in Europe for more than three centuries. The idea that projective geometry might check projectile physics never seems to have occurred to any thinkers besides Carnot in all that time. Everyone else just thought that harder-slung projectiles must be countered by thicker walls.

Lazare was still serving in France's Directory government in 1796, when his son, Sadi, was born. In his childhood, Sadi Carnot seemingly never knew a time without tumult. From 1796 to 1812, the Napoleonic Wars raged across the continent as Sadi grew into young adulthood. The year 1812 marked Napoleon's fatal invasion of Russia but also the first year of Sadi Carnot's formal education. Sadi entered the École Polytechnique that year, and, having been individually tutored for 16 years by his father, Sadi already had a profoundly shaped intellect.

It was 1820 before Sadi could disentangle himself from the military, and by then, he was 24 years of age. According to Mendoza, "The period that followed was the creative time of his life. He studied widely—at the Sorbonne, the College de France, the Ecole des Mines and elsewhere—concentrating on physics and economics" (ix). Sadi Carnot had the traditional engineer's faith in mathematics and believed that mathematical and economic theory could be applied to factory organization. Like many of the Renaissance men of the time, he read widely and found himself attracted to the arts.

Cross-curricular study combined with a formal education devoted to rigorous mechanical applications led Carnot to develop new insights. In 1823, the year of his father's death, Sadi wrote his own opus. Sadi Carnot wanted to write a work that could be understood by nontechnical readers, so he roped his younger brother into being a reader and editor. A year later, the book was published in Paris under the title *Reflections on the Motive Power of Fire, and on Machines Fitted to Develop That Power.* The first sentence states, "Every one knows that heat can produce motion. That it possesses vast motive-power no one can doubt, in these days when the steam-engine is everywhere so well known."

The connection between heat and motion, based on everyday observations, allowed Carnot to quickly express the universality of the phenomena:

> To heat also are due the vast movements which take place on the earth. It causes the agitation of the atmosphere, the ascension of clouds, the fall of rain and of meteors, the currents of water which channel the surface of the globe, and of which man has thus far employed but a small portion. Even earthquakes and volcanic eruptions are the result of heat.
>
> From this immense reservoir we may draw the moving force necessary for our purposes. Nature, in providing us with combustibles on all sides, has given us the power to produce, at all times and in all places, heat and the impelling power which is the result of it. To develop this power, to appropriate it to our uses, is the object of heat-engines. (3)

The connection between heat and motion blurs the cause and effect. A modern understanding of thermodynamics indicates, through the first law of the discipline, that "work creates heat." However, the dissipated heat can then be used to create work in another system. Carnot's slight error in the first sentence later gets encoded in the word "energy."

Again, Carnot's education as a practical mechanical engineer and soldier prevented him from flitting off into philosophy (if only Russell, Wittgenstein, and Gödel ever learned how to work on an engine, think of what they might have accomplished). He relates his insights to practical observations:

> The production of motive power is then due in steam-engines not to an actual consumption of caloric, but to its transformation from a warm body to a cold body, that is to its re-establishment of equilibrium—an equilibrium considered as destroyed by any cause whatever, by chemical action such as combustion, or by another other. We shall see shortly that this principle is applicable to any machine set in motion by heat.
>
> According to this principle, the production of heat alone is not sufficient to give birth to the impelling power: it is necessary that there should also be cold; without it, the heat would be useless. (7)

The steam engine was to Carnot what the cannon was to Newton, a machine that created a boundary condition (for motion in the first case and for heat in the second) that compelled each to see a universal principle. The above excerpt, like the first sentence of Carnot's work, contains an error that is somewhat difficult to see (and impossible to see in Carnot's era). Of course, heat always flows from hot to cold, but heat is a manifestation of entropy, and, as (again) will be shown in a later chapter, the first cause of the universe must be from cold to warmth if it is to operate within the modern understanding of thermodynamics.

Most significantly for quantum mechanics, Carnot laid the foundation for the noncommutative property:

> Let us imagine that . . . the piston having arrived at the position of *ef* [previously noted in the text], we cause it to return to the position *ik* [previously noted in the text], and that at the same time we keep the air

in contact with the body A. The caloric furnished by this body during the sixth period would return to its source, that is, to the body A, and the conditions would then become precisely the same way as they were at the end of the fifth period. If now we take away the body A, and if we cause the piston to move from *ik* to *cd*, the temperature of the air will diminish as many degrees as it increased during the fifth period, and will become that of the body B. We may evidently continue a series of operations the inverse of those already described. It is only necessary under the same circumstances to execute for each period a movement of dilation instead of a movement of compression, and reciprocally.

The result of these first operations has been the production of a certain quantity of motive power and the removal of caloric from the body A to the body B. The result of the inverse operations is the consumption of the motive power produced and the return of the caloric from the body B to the body A; so that these two series of operations annul each other, after a fashion, one neutralizing the other.

The impossibility of making the caloric produce a great quantity of motive power than that which we obtain from it by our first series of operations, is now easily proved. (18–19)

Although this section is wordy, Carnot effectively makes the case that all machine interactions result in a loss of heat. This is why it is impossible to get more out of an interaction than is put into it. The more interactions that exist between a source of heat and the work needing to be done, the more inefficient the machine. The less effective the conduit of the heat, the more inefficient the machine itself will be. Heat loss can be reduced in interactions but not eliminated.

Carnot could not have stated the principle like this because the knowledge did not exist at the time, but interactions at a molecular level continue all the time, meaning that even when not working, the machine itself radiates.

The irony in this is that a Carnot cycle now refers to a machine where heat is fully reversible. Nothing like this can exist, but the concept exists as a mythic exemplar for engineers to try to achieve. This is not foolhardy, as at some point, the question of whether one can get more juice out of less squeeze must be answered in a series of 9s to the right of a decimal, but the perfect Carnot cycle was disproved by its namesake.

Sadi Carnot died of cholera, at just 36 years of age, in 1832. His brilliant insights, as impressive as any ever thought, lay dormant for nearly two decades until rediscovered by a German physicist named Rudolf Clausius. But before we can understand Clausius, we must understand his historical antithesis: Lavoisier.

O

Lavoisier and the Law of Conservation

THE TRAIT OF INTELLECTUAL CURIOSITY, A PREREQUISITE FOR GENIUS, belongs to no social class, ethnic group, or religious affiliation. The intellectually curious sometimes find themselves in environments that foster the development of that curiosity and sometimes find themselves in environments that actively seek to destroy it.

The irony of the French Revolution is that, prior to 1789, the nobility lived under conditions that allowed for a class of "gentlemen philosophers" (to borrow a term more often used to describe Englishmen who were well-to-do enough to finance their scientific investigations) to engage in original research. The revolution, of course, destroyed this class of self-funded and self-taught researchers but then created the university institutions that allowed for the flourishing of the great minds mentioned in the previous section.

The tragedy is that one can only imagine the type of conversation that Antoine Lavoisier and Lazare Carnot might have had if politics had not rent French politics asunder. As it was, the two stand forever on opposite sides of a chasm. Lavoisier, born in 1743, hailed from the nobility. As such, he enjoyed access to a formal university education. He eventually attained a law degree and license to practice (perhaps only to please his father and grandfather, who were attorneys) but never actually took on any cases. It seems that he initially explored cross-curricular interests but narrowed his focus to the study of chemistry. Eventually, Lavoisier focused on scientific investigations through the development of ingenious experiments.

His first insight of historical significance was published in 1778 as he disproved the "phlogiston" theory. This theory evolved from the ancient elemental notion of matter: the belief that fire, water, and air (or just one of these) formed the basic composition of matter. Expounded in the 17th century by the scientists Johan Becher and Georg Stahl, the theory held that an element named phlogiston was common in all elements and that every chemical interaction led to the release of it in the form of flames, heat, and/or light. (Lavoisier recognized that oxygen was in fact needed for combustion to occur. However, the "phlogiston" theory would have been an interesting component to the work of Sadi Carnot.)

For the purposes of this narrative, Lavoisier made two cornerstone contributions to scientific/philosophical history. The first of these came in 1787 with the publication of the *Method de Nomenclature Chimique* (a dictionary/encyclopedia of chemistry nomenclature), which Lavoisier had written along with his fellow chemists Berthollet, Fourcroy, and Guyton. *The Elements of Chemistry*, as it is called in English, established the basic definitions for chemistry and also functioned as a handbook for how to conduct experiments in chemistry.

The establishment of a nomenclature, like the publication of Cardano's book on the use and theory of algebra, greatly advanced science in Western Europe while at the same time ossifying the science against future insights. Future quantum physicists, such as Niels Bohr, recognized that the concepts and truths of quantum physics could be applied to chemistry but that a wall of already established terms made such an integration difficult or impossible.

The second of Lavoisier's great discoveries came in the form of the law of conservation of mass. If one performs a closed experiment, the weight of the finished product will be the same as it was in the beginning. Other thinkers can lay claim to the founding of this insight, but no one had done more to prove the hypothesis with actual experimentation than Lavoisier (and he did it with the kind of beakers, burners, and lab equipment associated with classic novels like Robert Louis Stevenson's *Dr. Jekyll and Mr. Hyde*).

Eventually, the chemical symbology used to express the law of conservation of mass would come to be known as stoichiometry.

Stoichiometry is chemistry's version of the algebraic commutative property. The weight of chemical composition on the left side must be the same weight as the chemical composition on the right side. If, for example, you turn three potatoes into mashed potatoes, the weight should not change. The same is true if you combine sodium (Na) and chloride (Cl) into table salt (NaCl): the compound should weigh as much as the two original elements.

The connection between the law of conservation of mass and the law of conservation of energy would occur, interestingly enough, because of a later experiment done on human blood. In 1842, a German physician named Julius Robert von Mayer traveled to the Dutch East Indies (today's Indonesia) and was surprised to find that the blood in the veins of people living in such a tropical climate was redder than the blood in the veins of Europeans.

Mayer hypothesized that in a hot climate, less oxygen had to be burned to warm the human body. This meant that heat and work (a physics term meaning the transference of energy) must be the same force, eventually to be called energy. It is surprising to see a scientific insight of such broad application come from the study of blood—but not without a dark irony. After all, the founder of the law of conservation of mass lost his head at the height of the revolution's Reign of Terror. Lavoisier, a nobleman, was also one of the hated tax collectors. Arrested in 1793, the guillotine separated his head and body on May 8, 1794. The two sections of Lavoisier weighed the same after the execution as they did before it.

P

Rudolf Clausius, Energy, and Entropy

ONE OF THE REASONS THAT THERMODYNAMICS FAILED TO BECOME THE centerpiece of the Western scientific narrative must surely be the prose style of the discipline's major authors. For all of Sadi Carnot's brilliance, his central thesis gets dragged down in the use of examples involving hypothetical machines. The scientist and thinker who inherited, corrected, and clarified Carnot's work was Rudolf Clausius (1822–1888), and much of Clausius's genius is obscured by thickened prose. If either Carnot or Clausius could have written as clearly as Darwin (or even Newton, who employed a deliberately opaque style), then they might be seen as the two most important theorists in the history of science. The author will try to rescue Clausius from his writing in this section.

Born in Prussia, Clausius attained his university degree in mathematics and physics from the University of Berlin in 1844 and then earned a doctorate four years later from the prestigious University of Halle. By 1850, he taught as a professor at a military and engineering college at Berlin and did his most original work there.

Clausius does not possess a colorful personal history; he appears to be one of the first of the steely-eyed engineering specialists, so common in industry, who had little use for philosophy. He does not employ explanatory analogies in his writing and seemed to be interested only in understanding the direct mechanical applications of his work. Perhaps because of this and perhaps because he never produced any eerie equations or mind-bending revelations about time and space, history has missed marking him as the greatest genius in science.

In his 1850 work *On the Motive Power of Heat, and on the Laws Which Can Be Deduced from It for the Theory of Heat*, Clausius begins with the following:

> Since heat was first used as a motive power in the steam-engine, thereby suggesting from practice that a certain quantity of work may be treated as equivalent to the heat needed to produce it, it was natural to assume also in theory a definite relation between a quantity of heat and the work which in any possible way can be about the nature and the laws of heat itself. In fact, several fruitful investigations of this sort have already been made; yet I think that the subject is not yet exhausted, but on the other hand deserves the earnest attention of physicists, partly because serious objections can be raised to the conclusions that have already been reached, partly because other conclusions, which may readily be drawn and which will essentially contribute to the establishment and completion of the theory of heat, still remain entirely unnoticed or have not yet been stated with sufficient definiteness.
>
> The most important of the research here referred to was that of S. Carnot. . . . Carnot showed that whenever work is done by heat and no permanent change occurs in the condition of the working body, a certain quantity of heat passes from a hotter to a colder body. (109)

One question to be pondered is this: Does Carnot's assertion about heat passing from a hotter to a colder body count as the discovery of entropy and thus of the second law of thermodynamics?

Given that the universe around an interaction will be cooler than the heat generated by the motion of the interaction, this author concludes that Carnot discovered entropy even if he did not state it quite as explicitly. Others might disagree and credit Clausius. Others might think that the notion of pinning blue ribbons labeled "discoverer" on individuals who are clearly part of an intellectual process that gradually evolved is a pointless endeavor.

What Clausius actually added, in 1854, was that a change occurs when the heat transfers. The Wikipedia page on Clausius states this as a central quote of his work: "Heat can never pass from a colder to a warmer body without some other change, connected therewith, occurring at the same time."

Yet in 1850, Clausius made the same statement (albeit with more words) and did so by synthesizing the heat/work studies of Joule:

> It is . . . almost impossible to explain the heat produced by friction except as an increase in the quantity of heat. The careful investigations of Joule, in which heat is produced in several different ways by the application of mechanical work, have almost certainly proved not only the possibility of increasing the quantity of heat in any circumstances but also the law that the quantity of heat developed is proportional to the work expended in the operation. To this it must be added that other facts have lately become known which support the view, that heat is not a substance, but consists in a motion of the least parts of bodies. If this view is correct, it is admissible to apply to heat the general mechanical principle that a motion may be transformed into work, and in such a manner that the loss of vis viva is proportional to the work accomplished.
>
> These facts, with which Carnot also was well acquainted, and the importance of which he has expressly recognized, almost compel us to accept the equivalence between heat and work, on the modified hypothesis that the accomplishment of work requires not merely a change in the distribution of heat, but also an actual consumption of heat, and that, conversely, heat can be developed again by the expenditure of work. (110)

The final sentence of that section is the most important in science, but its significance has been clouded by the out-of-order nature of word development.

The term "energy" carries a positive connotation in the sense that it is the force that can create order. Students of biology learn of photosynthesis, where energy from the sun gets absorbed by plants and is then cycled through a system of life. In the vernacular, the word "energy" tends to be synonymous with "pep" and "positivity," and there are few ways to describe something or someone as "energetic" in a negative sense.

The word "energy" itself was first used and recorded in 1802 by a British scientist named Thomas Young. He intended the word to explain the synthesis of mass and velocity. Today, the term used by physicists to describe the same phenomenon would likely be "force," as F = MA stands for "force equals mass multiplied by acceleration" (velocity is acceleration with a direction). However, the word "energy" did not really enter into the scientific lexicon until almost the middle of the 19th century and became common before the definition of the word or its implications became well understood.

The fact that Clausius defined heat loss but then also participated in the shaping of the term "energy" indicates how difficult the concept was to define. The thinkers of the mid-19th century, for all their brilliance, set up a definitional muddle. Clausius worked with the Scottish physicist William John Macquorne Rankine (1820–1872), who specialized in the study of steam engines to develop a concept of energy. The DK book *Physics* (2012) states the situation thusly:

Clausius and Rankine worked on the concept of energy as a mathematical quantity—in the same way that Newton had revolutionized our understanding of gravity by simply looking at it as a universal mathematical rule, without actually describing how it works. They were finally able to banish the caloric idea of heat as a substance. Heat is energy, a capacity to do the work, and heat must therefore conform to another simple mathematical rule: the law of conservation of energy. This law shows that energy cannot be created or destroyed; it can only be transferred from one place to another or converted into other forms of energy. In simple terms, the first law of thermodynamics is the law of conservation of energy applied to heat and work. (88)

At the same time that Clausius and Rankine used the word "energy," the conceit also got connected to theories regarding temperature, and this led to a sudden and almost alchemical bonding of the definitions of "heat," "energy," "temperature," and "conservation." Ihde's *The Development of Modern Chemistry* (2012) contains these two crucial paragraphs, which serve to not only describe the scientific moment but also preserve the way in which the historical record was preserved:

William Thomson (1824–1907), better known as Lord Kelvin, was professor of natural philosophy at Glasgow from 1846 until 1899. In developing the consequences of Joule's doctrine of the interconvertibility of heat and mechanical work he arrived at the concept of absolute temperature in 1848. The observation that thermometers based on the expansion of various gases, liquids, or solids failed to agree led to the conclusion, based on a study of Carnot's theory, that perfect heat engines, operating between the same temperature limits, should be equally efficient regardless of the nature of the gas employed. He suggested that equal temperature increments on an absolute scale might be considered ranges over which a perfect heat engine would operate at equal efficiencies. In 1854, following the general abandonment of the caloric theory, Thomson proposed an absolute scale in which equal amounts of work are produced by equal temperature increments. This was shown to correspond reasonably well with the scale of the gas thermometer.

The term "energy" came into use in the mid-1800's. Some years earlier, Thomas Young had pointed out the confusion resulting from use of the terms "force" and vis viva by various investigators. The acceptance of "energy" in a precise scientific sense was largely due to Thomson and William Rankine. (399)

At roughly the same time, in 1855, Clausius introduced the Clausius theorem, or the concept of entropy. The word "entropy" was deliberately derived from the Greek word meaning "transformation." The underlying conceit was that when energy works through a machine, it moves from a higher to a colder temperature but that downward flow of temperature moves randomly but inevitably toward the cold. The cooling of the system leads eventually to a dissipation of the energy, and without outside energy, all systems fall into disorder. The way in which this happens might be random, but that *it will happen* is statistically ensured.

The concept of entropy gets confusing when applied to wildly different energy sources. A 19th-century engineer might concern himself with entropy only as a way of realizing that a single bin of coal will never be able to power a steam engine perpetually. Engineers can increase efficiency by ensuring that as much heat as possible moves directly from the coal to the heating of water, but the system will always tend toward entropy.

It is harder to see the concept at work with something like a star, where entropy is just as surely always at work, or with a black hole, where entropy occurs at the slowest known rate but always occurs. The confusion shows up various forms, but this part of the book deals only with how the confusion began; solutions to these problems will be offered in part II.

Q

James Clerk Maxwell

By 1860, the second law of thermodynamics, a concept first recognized by Sadi Carnot in 1824 and then formalized by Clausius, existed as a foundational scientific concept. That the second law of thermodynamics relates to the first concept (heat loss) that was discovered is fitting given the murky history of this field of study. Clausius and Lord Kelvin (aka William Thomson, but in this author's opinion, anyone who can go by the title of "Lord" should). The first law of thermodynamics evolved more than it was discovered but seems to have been established about the same time.

The first law of thermodynamics states that in a closed system (which does not allow matter to enter or escape), the energy used will be the equivalent of the work done. Stated more succinctly, work creates heat. The first law can be derived from any form of motion. People shiver in cold weather to try to create body heat. Any running machine generates heat. The presence of heat indicates life just as the absence of it indicates death. A dead body, from a thermodynamic perspective, is one where the cellular mechanisms of life have ceased to work.

The first law of thermodynamics commutes. If, in the 19th century, Clausius explained the first law to a mathematician, then that mathematician would have been able to apply Cardano's algebraic concepts to the transfer. If work creates heat, then heat must create work. This works out mathematically but can be disproved by any experiment. If you heat up water to produce steam and the steam moves a piston, then you should be able to push the work from the piston back into the water. It doesn't work that way, though. Heat things up all you want, and it won't necessarily create work.

To understand why this is, one must apply the second law of thermodynamics, which explains that every interaction, whether between cells, chemicals, or gases in the center of stars—every *single* interaction, *every one* that has ever happened or ever will happen—results in a loss of heat. That loss of heat, in some capacity, means the breakdown of bonds in a structure and the inevitable movement toward disorder.

The encoding of the first and second laws coincided almost exactly with the 1859 publication of Charles Darwin's *On the Origin of Species*, and this might explain why it was evolutionary theory, not the notion of a universe hopelessly careening toward disorder, that threw Victorian England into a tizzy. Unlike the French and German engineers, Darwin possessed a florid prose style and effectively used analogies.

Also, since he barely comprehended mathematics, Darwin felt no need to stack equations up between his readers and the comprehension of his theory. *On the Origin of Species* is as readable today as it was in 1859 and can be read as a landmark in both science and literature; it immediately sold out on publication, became the talk of the English-speaking world, and essentially created the field of modern biology.

At precisely the moment that Clausius and others came to understand that everything that moves loses heat, and that heat represents the loss of the bonds holding everything together, Darwin came to understand the progressive means by which organisms can capture "energy" from the environment for the purpose of creating order. One could conceive of an alternative history of science where Christian theologians might have embraced Darwin as a secular alternative to Clausius.

But it didn't play out that way. Writers and theorists have been analyzing the impact of evolutionary theory on religion and philosophy for years, but the two exemplars are A. N. Wilson and Daniel Dennett.

Wilson, in his book *God's Funeral: The Decline of Faith in Western Civilization* (1999), detailed the effects of evolutionary theory on Victorian intellectuals. They suddenly recognized that the biblical story of Genesis could be read only as an allegory or just as pure fiction.

Thinking people had been inclined to believe that for centuries, but without a clear alternative theory to describe the existence of life, many defaulted to a watery agnosticism. The arrival of evolutionary theory made it impossible to accept the truth of Genesis and therefore removed any reason for believing in the rest of the biblical narrative.

In his book *Darwin's Dangerous Idea* (1995), Daniel Dennett argues that Darwinian theory did more than just refute the Book of Genesis (many scientific theories or even common sense would do just as nicely). Evolutionary theory upended the basic conceit of creationism: that it takes a big and complicated thing to make a slightly less complicated thing. In short, a creationist would argue that it takes a human to make a watch. An evolutionary biologist would argue that the watch is not singular; it is part of a "bush" of timepieces that can trace its ancestral lineage back to water clocks and sundials.

On the Origin of Species drew both the attention of scholars and the fire of theologians for almost a century. This meant that the first and second laws remained shielded from religious scrutiny and, because they were seen as being mere components to mechanical engineering (dry stuff for dry people), not relevant to any theorizing involving the existence of God. When thermodynamics did finally capture the public's attention, it was only because of the development of the atomic bomb. No one seemed to see the acceleration of entropy as a threat to theology unless it was as just another entry in the lengthy "theodicy" column.

Ironically, modern creationists often invoke the second law of thermodynamics as a counter to evolutionary theory. The thinking goes like this: (1) engineers and cosmologists tell us that the second law trumps everything. (2) But life is ordered. (3) Therefore, evolutionary theory violates the second law of thermodynamics and cannot be true. The implication here is that only God could be powerful enough to push back entropy and create life.

It's a thoughtless claim for a number of reasons. First, the second law works only on closed systems where no energy comes in from the outside. Earth is an open system, and we get our energy from the sun. Second, the implied God would also be subjected to the same second law, which means that God radiates and tends toward disorder. Third, the sun isn't "shining"; rather, it is dissolving, and life on Earth captures that temporary radiation to make temporary order. Everything dissolves and dies. The second law will not refute evolutionary theory, and it will not make you feel better about your place in the universe, but it is inevitably true.

R

The Curies and a Demon

WERE MARIE CURIE (1867–1934) AND HER HUSBAND PIERRE CURIE (1859–1906) chemists or quantum physicists? Do they trace their intellectual origins to Lavoisier or to the Carnots? Much of the Curies' work developed from seeing an analogous relationship between X-rays, which had been discovered in 1895, and the rays that uranium salts emitted. Uranium salts radiated without needing to absorb energy from anything outside, and this indicated an internal process. That internal process became the foundation of the work done by the Curies, which was, at that time, considered to advance chemistry.

The reason they were seen to be advancing chemistry is because that field, because of Lavoisier, had been established for about a century, while thermodynamics remained nebulous, even by the time of the Curies. If thermodynamics had been better understood, then the Curies would have been recognized for establishing new and exciting proof for the universality of the first and second laws.

Bananas go from green to yellow to brown to black, but they don't go back. Bananas are composed of weak molecular structures that break down quickly, losing heat and trending to disorder as they do so. The same concept can be applied to elements; however, because the molecular structures are stronger, the loss of heat and the trend toward entropy occur over vastly longer time periods. Both a banana and a bar of steel are rotting, but you can watch it happen only to the banana.

Some elements "rot" faster than others, which means they radiate. This radiation eventually "transforms" (the root word of "entropy") into another substance. However, the Curies performed experiments in an era before a theoretical framework had been constructed. This meant, unfortunately, that they also conducted their research experiments at a time before the impact of radiation on DNA (not yet discovered) was understood.

We now know that life on Earth, including humans, evolved to absorb only a particular amount of radiation. When one object captures the radiation of another object for the purpose of making temporary order, that is energy. Untold numbers of hydrogen interactions occur in the sun, and each of those interactions results in heat loss. This means that the sun radiates and that the radiation shoots out into the solar system, where most of it hits nothing, but a small percentage will shine through to Earth.

Earth possesses a magnetosphere, which shields out some of the most harmful aspects of the sun's radiation. This means that Earth gets just enough radiation (light) to provide the energy for life to exist. Life, however, can tolerate only a very particular amount of radiation—just what it evolved over hundreds of millions of years to use. Just as there is no such thing as a "poison" (two aspirin are medicine, 50 are lethal), radiation can come in helpful or harmful doses.

This explains why humans should not venture too far above or below the surface of Earth. If we go too high, then we get hit with an above-normal level of space radiation, which is a particular danger for astronauts living for any length of time in a space station. If we go below the surface of Earth, we encounter metals that might radiate too quickly for our DNA to handle, and this means that the replicator functions in our cells can get torn apart, resulting in cancerous growths.

The Curies had no access to any of this as-of-yet-undiscovered and -uncompiled knowledge, so they worked directly with uranium, which almost certainly led to some arguable level of radiation poisoning for both of them, but the story of their lives and marriage is well known and better told by other authors. In terms of their work, both saw the immediate analogy between uranium's radiation and X-rays.

X-rays emit a very particular level of radiation, one that is capable of penetrating the outer layer of human flesh. Bones are dense enough to absorb the radiation, which then shows the positions of the bones. It's sort of like throwing sand into an empty room in order to find the Invisible Man. Marie Curie would later, after the death of Pierre, put her knowledge of X-rays to practical use by designing a mobile X-ray machine and using her reputation to gain funding for their production and use.

Eventually, the work on uranium led Marie to understand that movement in the nucleus led to the release of energy that came in the form of radiation. This could mean only that some elements contain a central structure that is unstable and would in time lead to the understanding that all elements can be classified by how much movement occurs in the core of the atom. Elements with little movement are stable, and those with a lot of movement (and thus a lot of heat loss) are unstable.

In the same year as Marie Curie's birth, just as these fundamental understandings about the nucleus of elements were emerging, Plato rose from the grave and tried to close a door on the progress of scientific understanding. Plato didn't rise by himself; he was conjured up by James Clerk Maxwell (1831–1879), a Scottish physicist who applied kinetic theories of energy to gases, among many other contributions.

In an attempt to potentially counter the second law of thermodynamics, Maxwell posited a "being who can play a game of skill with molecules." Maxwell's thought experiment put this "being" at the center of separated chambers filled with gas. The "being" controls a door without mass and lets only quickly moving molecules through going one direction and only slowly moving molecules through going in the opposite direction. Since temperature is a fundamental measure of the speed of the molecules, this would allow for a swap of temperature without the loss of heat.

Lord Kelvin referred to the "being" who controls the door as a "demon," and the entire thought experiment eventually would be recorded in the history of philosophy and science by the exciting title of "Maxwell's Demon." This is the second-best example of how sexy phrasing can derail a serious scientific dialogue; the first and best example can be attributed to the physicist John Wheeler, who offhandedly renamed "completely collapsed objects" as "black holes" while he spoke at a routine physics conference in the 1960s, thus spawning a whole industry.

The problem, of course, is with the idea of both the demon and the "massless" door, both of which can come only from the pure Platonic realm of forms. An actual door would contain atoms and fields, which would interact with the particles and create heat loss. Since demons don't exist as actual animals, let's replace it with, say, a koala bear. The koala contains cells that interact and radiate. We have to include the door and the koala within the system, which means that as they trend toward entropy, so too, statistically, does the entire system.

None of that even mentions the box that the experiment occurs in. It radiates too.

This might see immaterial (pun intended), but other thinkers will put their hands over the periodic table and try to Ouija up various Platonic demons to challenge or clarify the movements of quantum physics. Erwin Schrödinger tried it in 1935 (more on this in the chapter "Uncertainty and Exclusion"), and the Otherworld coughed a dead cat right onto the board. Another cat, alarmed but alive, appeared at the same time. Modern cosmologists Ouija up a Platonic sun and Earth every time they want to make the reference point of a "year" when trying to calculate the age of the universe.

S

Maxwell and Electromagnetism

JAMES CLERK MAXWELL CREATED THE CONCEPT OF STATISTICAL fields, and he possessed the mathematical acumen necessary to support his theories in the kinds of dense theoretical constructs admired by mechanical engineers. However, in his book *Convergence: The Idea at the Heart of Science* (2017), Peter Watson wrote about a seeming paradox:

> [Maxwell] read the paper by Clausius about the diffusion of gases. The problem, which several people had pointed out, was that, to explain the pressure of gases at normal temperatures, the molecules would have to move very fast—several hundred meters a second, as Joule had calculated. Why then do smells—of perfume, say—spread relatively slowly about a room? Clausius proposed that each molecule undergoes an enormous number of collisions, so that it is forever changing direction—to carry a smell across a room the molecule(s) would actually have to travel several kilometers. (38)

While this section shows that Clausius discovered or envisioned the famous "drunkards's walk" statistical pattern that molecules take as they interact and interfere with each other while moving in a particular direction, Clausius's speculation, combined with the seeming paradox, led to a thought that would become the foundation for quantum physics:

> Maxwell saw that what was needed was a way of representing many motions in a single equation, a statistical law. He devised one that said nothing about individual molecules but accounted for the proportion

that had the velocities within any given range. This was the first-ever statistical law in physics, and the distribution of velocities turned out to be bell-shaped, the familiar normal distribution of populations about a mean. But its shape varied with the temperature—the hotter the gas, the flatter the curve and the wider the bell. (39)

Work creates heat, and the hotter something is, the more likely it is to have a significant standard deviation from the mean. As you heat something up, the specific behavior of individual particles/radiation packets becomes increasingly less predictable. This is partially why radioactivity is so dangerous; thermonuclear weapons are subject to probabilities in the sense that a properly working weapon still might not work part of the time if the pathway of the initial charge fails to create fission. This also means that an unlikely pathway might be found and create fission where it was not intended.

A safety inspector for a nuclear weapon might perform an inspection on an electronic control board and find, statistically, that it is nearly impossible for an electric signal to be accidentally sent through the board. However, the introduction of extreme heat could melt the board and create new pathways (outliers) for a firing signal. Erich Schlosser pointed this out in his excellent book *Command and Control: Nuclear Weapons, the Damascus Accident, and the Illusion of Safety* (2013), about the safety hazards presented by nuclear weapons.

In any situation, micro or macro, the sudden introduction of heat will create greater deviations from the mean. From the quantum perspective, an initial movement that creates heat also stretches the probability field so that extreme outliers become more possible. This is why something really rare, like life, is more likely to exist next to a massive heat source, like a star. From a probability perspective, cold objects create a statistical line, with all waves and particles adhering to the mean. Movement generates heat, and the ends of the line melt away from the mean.

The closer something gets to a star, the more likely that unlikely behavior becomes—but only from the perspective of an individual particle. The "sweet spot" for life formation cannot, by statistical definition, be the most unlikely condition. Life cannot be rarer, say, than faster-than-light travel. The statistical distance between life and faster-than-light travel, however, is so vast that in order to have the heat energy to get near faster-than-light travel, you would burn up the statistical pocket for life's existence in the process.

Suddenly aware of a concept like a wave, Maxwell published 20 equations in 1861. This all came almost immediately after Clausius published his work on heat loss. Maxwell's equations synthesized electricity and magnetism into an electromagnetic theory, a concept as important for physics as Einstein's later insight about energy being analogous to mass. Maxwell, unfortunately, did not have a single trendy-looking equation to showcase his idea (*The Twilight Zone* would never, for example, feature those 20 equations floating during their introduction as the show did with $E = mc^2$).

Eventually, Maxwell shortened his equations to four, and in 1885, a British mathematician named Oliver Heaviside (1850–1925), who happened to be rather skinny, summarized Maxwell's concepts in easier-to-understand equations. The mathematical arrival of Maxwell's work more or less evolved with the development of equations that explained how Michael Faraday's (1791–1867) conceptualization of electromagnetic fields in 1831 could be explained with statistical certainty.

The creation of probability fields around phenomena provided a way to predict the motion of radiation but was dependent on the law of large numbers. Later, the marriage of quantum theory and probability would make it nearly impossible to predict where a quantum of radiation might end up once it is radiated because the hotter an object becomes, the more likely it is to become a statistical outlier. Even worse, as electrons scatter under conditions of extreme heat, they maintain an informational relationship with the other electrons they once resided with in an orbital shell. No matter where they go, one electron will spin one way, and the other will spin another way. If you know that two electrons were once in the same shell, then you will always know their relationship.

The only mathematical way out of the mess is to collapse a single particle into undefinable strings that will act with some sense of statistical certainty.

T

The Pre–Quantum Mathematics Crisis

WE CAN NOW SEE THAT THE COMMUTATIVE PROPERTY OF ALGEBRA throbbed like a pulmonary aneurysm in the heart of the mathematical certainty, but the potential rupture was not yet apparent because even as the 19th century became the 20th century, few theorists understood the radioactive properties of the elements. Confounding matters further, the 1869 creation of the periodic table by Mendeleev arranged the elements by atomic weight and their periodically appearing similar properties. The periodic table is a brilliant form of cataloging, but it supports the notion that elements are unchanging and defined as if their current properties are fixed. It likely ossified chemistry against concepts such as radioactivity defined by probability that would threaten a fixed picture.

Mathematics found itself challenged by potential internal contradictions, or paradoxes. Paradoxes are kinks in a line of logic, and they represent not a problem with anything in nature but rather an issue in our understanding of nature. In other words, paradoxes do not exist; they just indicate a problem with a theory. Once the theory is straightened out, the kink disappears. The commutative property seemed not to be a problem for mathematicians, but they were deeply troubled by the axiom and infinity.

In his classic *Makers of Mathematics* (2011), Stuart Hollingdale writes,

> By the beginning of the nineteenth century a clear distinction had been established between analysis, the study of infinite processes,

and algebra, which deals with operations on discrete entities such as the natural numbers. A major objective of much nineteenth-century mathematical effort was to unify—or, at any rate, to build bridges between—these two branches of mathematics. This endeavor was termed "the arithmetization of analysis." It was realized that the prime task was to construct a sound logical foundation for the real number system. Although the basic concepts of analysis—function, continuity, limit, convergence, infinity and so on—were progressively clarified and refined during the first half of the nineteenth century . . . much remained to be done. (353)

Pure mathematicians can sometimes be a little quirky. It's not clear that spending too much time with abstractions causes mental instability or that people with severe depression sometimes turn to mathematics as a cure. Mathematical thinking may be the most intense form of thought that humans can engage in and therefore crowds negative and persistent thoughts out of the neurons. (Bertrand Russell seemed to think this to be true, as he credited mathematics for saving his life during severe depressive episodes in his youth.)

This is relevant because we now veer away from the hard-eyed mechanical engineers like Clausius, Rutherford, and Kelvin, where theories are formed with fire, steam, and iron, and into a more lacily delicate world of mathematical theorems. The two great 19th-century thinkers were Richard Dedekind (1831–1916) and Georg Cantor (1845–1918). (It's admittedly weird that Cantor's last name is an anagram for "Carnot," but this is not a book of astrology, so it's enough to just note the coincidence.)

Dedekind had been a pupil of Karl Friedrich Gauss, the legendary German mathematician who generated the formulas for arithmetic and geometric sequences. Useless in the age of computers and a form of logic useful only for internal and "pure" mathematical purposes at the time, the equations are nonetheless still taught in mathematics classrooms and even tested in college entrance exams. Gauss's work, though impressive, should be considered tautological. It was true because math said it was true.

Yet tautological answers troubled Dedekind. Hollingdale writes, "He had realized as early as 1858 that the number system of real numbers lacked a firm logical foundation: the validity of such a simple statement as (square root of 2) × (square root of 3) = (square root of 6), for example, had never been rigorously established" (355). Mathematically, the issue has to do with the axiom. The first truth of a mathematical proof must prove itself or be proof of itself for the rest of the equation or argument to hold up.

A nonmathematical argument might be this: (1) It is true that water is wet. (2) Therefore, anything in water is also wet. (3) This sponge is in water. (4) Therefore, the sponge is wet.

One might ask if the word "wet" is an adjective for a property of water—a part of water's definition itself in the same way that "soft" might be part of the definition of "cushion" (to be harder would give it a function other than cushioning), but at some point, one will just have to concede some level of ambiguity and accept that "wetness," however defined, has something to do at least with water.

This next section is quoted at length because Hollingdale writes so well and makes impressive use of Dedekind's original text:

> We have seen how the Greeks, with their geometrical predilections, identified real numbers—first rational and, later, also irrational numbers—with line segments. They tacitly assumed that any such number could be represented by a unique point on an infinitely extended straight line with a specified origin. The rational numbers presented no difficulty; the problem lay with the irrationals. We can best convey Dedekind's approach by quoting from his 1872 essay:
>
> *The straight line is indefinitely richer in point-individuals than the domain of rational numbers is in number-individuals. The comparison of the domain of rational numbers with a straight line has led to the recognition of the existence of gaps, of a certain incompleteness or discontinuity in the former; while we ascribe to the straight line completeness, absence of gaps or continuity. Wherein does this continuity consist? Everything must depend on the answer to this question.*

He goes on to tell the reader how, after much pondering, he found the answer—what he calls the essence of continuity—in the following principle:

If all points of a straight line fall into two classes, so that every point of the first class lies to the left of every point of the second class, then there exists one and only one point which produces this division of all points into two classes, this severing of the straight line into two portions.

Such a severance, or partition, is now known as the *Dedekind cut*. He goes on to remark that this proposition may be taken to command universal assent: it is one of those truths that may be held to be self-evident. This is just as well, for he admits that he is:

Utterly unable to adduce any proof of its correctness, nor has anyone else the power. The assumption of this property of the line is nothing else than an axiom by which we attribute to the line its continuity, by which we define its continuity.

A more formal statement of this axiom, now known as the Dedekind-Cantor axiom, is that the points on a line can be put into one-to-one correspondence with real numbers. That is, to any point on the line can be assigned a unique real number, and conversely, any real number can be uniquely represented by a point on a line. (355–56)

Draw 10 points on a line, arbitrarily choose one to draw a line through, and then call all the points on the left "points on the left" and all the points on the right "points on the right," and what you will have done is create an arbitrary definition for "left" and "right" where the line is an axiom, or a self-evident truth.

The immediate analogy between the "left and right" part of a line is with the notion of negative and positive numbers. How can a number's positive or negative nature be defined based on an arbitrarily placed zero? Mathematicians are forced to admit that a number line is nothing but an axiomatic form of logic. If that point could not be refuted, then all these mathematicians were just shooting arrows into barns and then rushing in to paint bull's-eyes around wherever the arrow stuck.

The notion of a number line led to the conceit of infinity, which vexed Cantor, and the concept of infinity creates conceptual problems. The decimal points between 1 and 2 might be said to go on infinitely. How can both the space between 1 and 2 and the space between 0 and the "number that never stops growing" be infinite? You can see the problem.

Cantor gets unwarranted credit for stating in 1874 that size variations must exist within infinities (it was shown earlier that the medieval theologian and philosopher Robert Grossetest had already seen the paradoxes). Nonetheless, the modern problems with infinity can be traced to Cantor.

A variety of thought experiments can be derived from this concept, including the "Paradox of the Grand Hotel," which was conjured up in 1924 by the German mathematician David Hilbert. Hilbert imagined a hotel with infinite rooms where, somehow, every room is full. Two hypothetical guests show up, and while no new rooms are available, it is possible to shift guests into a new room in order to open one up. This is possible because there are infinite rooms, which means that rooms will be open to infinity.

Cantor's descriptions of the problems of infinity run into the opaque, and this is a situation where the problem can be understood only through the explanation of the solution. The concept of infinity is irrelevant in discussions about actual physics because infinity, as the concept is generally defined, simply fails to exist. If someone draws a line on paper and the pen runs out of ink, then the line on the paper is the extent of the infinity. One can theorize that with more ink, one could draw a longer line, but if you don't have more ink, then your infinity is where the pen went dry. The guiding analogy that clears up the confusion is that "infinity is like memory capacity."

If something exists, it cannot just be in a Platonic plane; it must be made of some material. In practice, for information to exist, it must be stored in something material, and this limits its capacity and creates two kinds of infinity. The first should be defined as theoretical infinity (TI), which is defined in the way that people typically think of "infinity": as a theoretical concept regarding a space or a number without any definite boundary or something that goes on forever.

The second type of infinity is real infinity (RI), which is defined as memory capacity. RI is what you have, and TI is what you could potentially have if you increased that memory capacity. This is why there is so much promise in the form of quantum computing. Computers that once stored information in vacuum tubes, then transistors, and then microchips could one day directly store information in atomic structures, or quantum bits.

For an example of the difference between TI and RI, we can look at pi. Computers can calculate pi out to 2.7 trillion digits. That 2.7 trillionth digit is the boundary of pi's RI. If every digit of pi is a Hilbert hotel room, then shifting one guest in the first room would push someone in a (very) distant room out. The same concept applies to prime numbers. The highest known prime number is 23,249,425 digits long. The last digit forms the RI boundary for prime numbers, and while we can conceive of the idea that another prime number might be found, the conceptual number that is the TI boundary for prime numbers does not really exist. This is all predicated on the axiom that prime numbers "exist" at all except as tautological mathematical concepts derived from axiomatic definitions.

Here's the thing: even TI is not really "infinite" in the accepted, Platonic sense of the word. If we stay with the prime number example, we can see that even TI must obey limits. If infinity equals memory capacity, then TI ends with the memory capacity of the cognitively capable beings in the universe. Computers and all forms of memory storage, in this example, should be seen as what Richard Dawkins calls an "extended phenotype," or a nonbiological manifestation of our memory capacities, in the same way that a dam is an extension of a beaver's genetic instinct to build a dam.

If all the memory capacity that humans have created could be used just to calculate a new digit of pi, then that would shrink all the other infinities, defined as non-prime numbers and the infinities between the prime numbers, to a single whole digit. This would happen because all the memory capacity would be used up in treating those as whole numbers. One could then increase the size of the TIs in between the prime numbers by moving back from 23,249,426 to 23,249,425 digits of pi, but the infinities would be limited by the memory capacity.

This example gets fuzzier the closer it gets to the boundary condition, but this is just an extension of the fact that if you have only a certain amount of ink, then you can either use that ink writing as many whole numbers as possible or use it writing decimal points in between whole numbers. If you increase one, then you shrink the other.

In studying this time period of mathematics, it's possible to feel the unease that the axiom and infinity brought to mathematicians. No one wanted to downgrade these elegant mathematical theories to mere models of physical interactions. Whereas at one time medieval scholars referred to theology as the "queen of the sciences" in the Catholic Church's educational structure, mathematics now reigned with the same title in secular universities. Mathematicians could no more renounce the Platonic realm than St. Thomas Aquinas could use logic to disprove the existence of heaven.

This explains why Bertrand Russell (1872–1970) and Alfred North Whitehead (1861–1947) embarked on writing the grandly titled *Principia Mathematica*. The title of their multivolume work deliberately paid homage to Newton's foundational work from 1687 but also indicated a deep ambition. Russell and Whitehead hoped to provide the same type of theoretical foundations for mathematics that Newton had for physics.

A psychological analysis of these two authors is unnecessary, but it might be noted that Russell lost both his parents before he could properly form memories of them. He was raised by aristocratic grandparents before they too died in his youth. It's likely that Russell's extraordinary intelligence would have alienated him from society even without these tragedies, but as it was, he seemed to find that only the intense study of mathematics could prevent him from manifesting his depression in suicide. There may have been a very worried part of him that believed that his sanity depended on the foundations of mathematics, and if those foundations broke, then what?

The authors wanted to prove that mathematics was more than just a series of tautologies—that a good foundational reason could be made for dividing a number line into negative and positive numbers—and therefore set theory was more than just an arbitrary classification system. They would, through their writing, intuit the Platonic realm and make it as real as a cannonball's elliptical orbit.

After years of toil, the authors found a reluctant but willing publisher, and their work came out in more than three volumes between 1910 and 1913. It famously took the authors about 360 pages to establish the logical foundations behind the equation $1 + 1 = 2$. The entirety of *Principia Mathematica* might obviously be compared to the quest for a Holy Grail in the sense that the Grail does not exist, and even if you found it, you would not be able to verify it as the true Grail, but the journey itself might teach the searchers something. In the Arthurian legends, the Grail hunters met around a round table, but the searchers for the mathematical Grail would eventually gather around a Vienna Circle.

U

Einstein

John Gribbin's book *Einstein's Masterwork: 1915 and the General Theory of Relativity* (2016) shines new light not only on relativity theory but also on Einstein's intellectual processes. Einstein's work fits well into a narrative of scientific history already established by Gribbin in his earlier works: *Schrodinger's Kittens and the Search for Reality* (1996) and *The Scientists: A History of Science Told through the Lives of Its Greatest Inventors* (2003). Interestingly, in this later work, Gribbin stakes out a position of historical determinism, one that argues that individuals usually do not matter that much:

> What is much more important than human genius is the development of technology, and it is no surprise that the start of the scientific revolution "coincides" with the development of the telescope and microscope. I can think of only one partial exception to this situation, and even there I would qualify the exception more than most historians of science do. Isaac Newton was clearly something of a special case. (xix)

Keep this observation in mind as we consider that Isaac Newton is the only thinker to whom Einstein can be compared in terms of historical significance, and Gribbin wrote that statement in 2002. His insights in *Einstein's Masterwork* should be understood in light of what he wrote then. He is a believer in studying the historical conditions of a person's time for putting the impact of the work into context. For example, "What made Newton and Einstein so special was that they didn't have just one brilliant idea (like, say, Charles Darwin and his theory of natural selection) but a whole variety of brilliant ideas, within a few months of one another" (4). One should read this not as a physicist defending the preeminence of his subject against biology but rather as an overall historical determinist making the case that if there ever were two people whose contributions can be explained only through individual genius, then it would be Newton and Einstein—and for the reasons stated.

As a young man in his early twenties, Einstein enjoyed socializing over coffee in the Swiss city of Zürich. The notion of atoms as fundamental particles, Gribbin writes, had yet to be established. The desire to prove that atoms were composed of more than pure theory called Einstein out into the realm of philosophical science. "This was what appealed to Einstein; the idea that the power of the human mind and mathematics was alone enough to conjure up deep truths about the world" (25).

One wonders if perhaps Einstein's social success in the cafés amounted to a phase. Newton, being the sort of man who stuck sewing needles under his eyeballs and liked (probably) fooling people into believing that fallen fruit featured at the center of his greatest insight, never married and never produced children. No one suffered from his detachments even if money cheats did later suffer from his work as a top bureaucrat at the royal mint. Einstein may have understood everything that Newton had scientifically, but Newton seemed to know something that Einstein did not: you do not get to take anyone with you when you go out there.

By this time, the problem that Einstein would need to solve had been established by science in 1901. Gribbin writes,

> The German physicist Philipp Lenard had discovered that electrons could be knocked out of the surface of the metal by shining ultraviolet light onto it. Strangely, he had also discovered that the energy of the electrons ejected from the metal surface did not depend on how bright the light was. Whether it was faint or dim the electrons always came out with the same energy (essentially, the same speed). How could this be? Just four years later, Einstein would explain the phenomenon. (33)

Einstein's Masterwork then enters the Annus Mirabilis phase of Einstein's life, and Gribbin lays out the four great works of Einstein's miracle year. Gribbin begins with the doctoral thesis, a pedestrian work (for Einstein) that "wouldn't tax the imaginations of the professors at the university too much" (47–48). Einstein produced no new research for his doctoral work; instead, he developed a mathematical system for understanding how sugar molecules would change sizes by absorbing water in a solution and then further used these equations to determine how to measure the solution's viscosity. The dissertation was unimpressive only in comparison to Einstein's other work, which is to say that it didn't alter a paradigm and warp everyone's mind, but "he developed techniques with widespread applications for industry wherever suspensions of particles in liquids are used" (51). The work simplified a process.

Most important, Einstein's dissertation prepared his mind to think in terms of particles and their behaviors. Atoms are fundamental to molecular theory, and thinking like this likely prepared him to think of light as being made of photons rather than waves. Gribbin writes,

> In the paper that became his doctoral dissertation Einstein had already used the idea that molecules of sugar dissolved in water are being bombarded by water molecules from all sides, and that the way the sugar molecules move through the sea of water molecules affects the measurable properties of the solution: its viscosity and its osmotic pressure. The success of Einstein's results provided powerful circumstantial evidence in favour of the kinetic theory, but even that was not direct proof that atoms and molecules exist. (61)

This is the second paper of 1905 that focused on what is known as Brownian motion, which Einstein statistically predicted. Even the use of circumstantial evidence from that paper may have had its important effects; Einstein was about to engage in a process of inverse questioning. A phenomenon now existed that required explanation; this was Lenard's puzzle regarding the fact that the intensification of a light beam did not increase the ejection of electrons from the object that received the light.

Particles of size that are hit by molecules will usually take the hit relatively equally on all sides, but constant bombardment produces rapid statistical anomalies, which means that occasionally the force of the impact will favor one side over the other, causing a large particle to move in herky-jerky motion but nonetheless in a general direction. This is the famous "drunkard's walk" concept of particle motion, but it can be determined only with statistical models.

By the time Einstein got to Lenard's puzzle regarding the intensity of light and electron discharge, he had been thinking in terms of particles, statistics, and inverse questions for some time. Perhaps equally as important, he had built up confidence in his methods. Here, Einstein would need to connect to the mind of his intellectual antecedent. Gribbin writes of the situation regarding illumination at that time:

> Isaac Newton thought of light as a stream of tiny particles, and used this model in his attempts to explain his observations. . . . His 17th century Dutch contemporary, Christiaan Huygens, had argued for a different interpretation of the same phenomena, based on the idea that light is a form of wave; but Newton's model held sway (largely because of the god-like status that his successors gave to Newton) until the work of the Englishman Thomas Young and the Frenchman Augustin Fresnel early in the 19th century. (69)

Young developed the famous double-slit experiment, which indicates that waves of light cancel each other out through a rippling and distinctly wavelike motion. These experiments, as it turned out, did not establish light as a wave. Instead, the experiment revealed a property of light that could be exposed in that instance and that could be described only by thinking of light as a wave. Einstein understood that in order to solve Lenard's puzzle, he would need to think of light in a different way.

Max Planck established the starting point of a revolutionary syllogism with a 1900 paper that stated that light could be thought of as quanta. Blinded by waves, Planck did not take the syllogism to its natural conclusion. In 1905, Einstein theorized that light acts in packets of energy—particles—and used the photoelectric effect as proof. This is one of the most interesting parts of Gribbin's book because he makes it clear that although Einstein would later win the Nobel Prize for the "photoelectric paper" (as it is now known), in fact "the section on the photoelectric effect was only a relatively small part of the paper: one of several examples that Einstein used to illustrate the importance of the concept of light quanta" (82).

Light, thought Einstein, looked like a wave only statistically. Particles could present as statistical waves at a certain level. This is the first inclination that the "observer" could be fooled relative to the stance of observation. This freed Einstein to theorize that intensifying a light beam did not increase the power of waves but instead increased the number of photons. This would solve Lenard's puzzle. More photons would knock off more electrons but not increase their speed. A greater intensity of waves, producing different colors, would increase the overall energy shooting into the electrons and therefore increase the speed. (The Nobel Prize came for Einstein in 1922 and evidentiary proof for his theory in 1923.)

But 1905 was still not over, and Einstein began to build on the work of other theorists who were working on the concept of motion and how it related to an observer. Einstein, having already broken with the wave theory to produce a model that better explained the photoelectric phenomenon, now broke away from the concept of the ether because "he said that what matters is how two objects move relative to each other, and that there is no absolute standard of rest against which motion can be measured" (100).

This is the central concept of the fourth paper, the one that features special relativity. Einstein started with new first principles, referencing no one, and began with the principle that some aspects of physical reality are purely related to the relation of two objects. The "ether" might be described as a stable reference by which motion could be measured, but Einstein understood that two objects traveling at the same speed—or a force like electricity/magnetism that moves through two conductors that are the same speed—provide their own reference and that movement is relative to the objects and not to the ether. This made two points for understanding motion rather than three. It is the difference between standing on a rock and watching a rabbit racing a turtle and standing on the turtle while watching a rabbit race a turtle. Gribbin writes,

> Einstein proved that an observer in one inertial frame would perceive objects in a different inertial frame shrunk in the direction of their motion relative to him, and he would see clocks in the other inertial frame running more slowly than clocks in his own inertial frame. An observer in the other inertial frame would see the mirror image of this—he would see the first observer's clocks running slow, and the first observer's rulers and other equipment (and, indeed, the first observer) shrunk. All of this has now been confirmed by experiments. (108)

This work led to the understanding that mass and energy are the same thing, thus $E = mc^2$. This 1905 culmination was the special theory of relativity, which might be better described as a specific theory because it applied only to the phenomena of energy and mass. Could it be applied more generally to explain a larger picture? It's the equivalent of Darwin realizing that natural selection applied specifically (specially) to beetles and finches but wondering whether it applied to all life everywhere. Gribbin states that the outpouring of work by Einstein in 1905 qualified him as a genius but "was ahead of its time." Einstein stacked one brilliant insight on top of another in 1905, but he was really finishing syllogisms started by other physicists. The year 1915 would be different.

Mathematicians scribbled in the margins of Einstein's theory, providing a greater validity for his ideas but also piquing the theorist. Few people knew what he was talking about, so honors and positions came slowly. He kept up his work at the patent office but started looking for new jobs. In the funniest section of the book, Gribbin writes that Einstein

> actually applied for a post teaching mathematics at a Zurich high school, enclosing with his application copies of all his published scientific papers. The bemused school governors did not shortlist him for an interview, even though in his covering letter he pointed out that he would be able to teach physics as well. How different might the development of physics have been if they had had more imagination? (119–20)

By 1909, Einstein had secured a professor's post that paid him the same as what he made in the patent office, but "rather than settling down in Zurich and working his way up the academic ladder there, Einstein's move marked the beginning of his years as a peripatetic professor, hopping from university to university in search of . . . the opportunity to work on what he wanted" (123).

Einstein's insights into motion and gravity began by understanding that "a theory of accelerated objects is also a theory of gravity. The explanation is disarmingly simple, although nobody before Einstein spotted it" (135). Einstein used an analogy with elevators, but Gribbin updates it to spaceships. Because gravity and acceleration are indiscernible, once the rockets on a ship go off, everything starts to float. Shine a light on the wall at that time, and it will make a little bright spot. Then turn the rocket on again so that the ship shoots forward, and "everything falls to the floor. . . . But what happens to the light beam?" (136).

Light moves at a speed, so while the photons are en route, the ship moves forward a little, and the beam will hit a different spot on the ship. Observers will see a bent beam of light. This means that "gravity must bend light by just the right amount to make acceleration and gravity precisely equivalent" (136). He published in 1911 and made this portion of the theory subject to evidence collection. An experiment was set up in 1914, but World War I got in the way. This was good, writes Gribbin, because Einstein's math predicted the wrong result at that point. It should enhance our conception of Einstein's genius to know that he was smart enough to go find help when he needed it. He got a numbers man, Marcel Grossmann, to help with the equations.

From "First Steps," Gribbin moves into "What Einstein Should Have Known," which features geometric concepts that had been worked out since the 18th and 19th centuries. Carl Friedrich Gauss created non-Euclidean geometry (dealing with other than flat surfaces) in the late 18th century but did not share it with anyone. Other geometers found the same concept on their own slightly later:

> They all hit on essentially the same kind of "new" geometry, which applies on what is known as a "hyperbolic" surface, which is shaped like

a saddle, or a mountain pass. On such a curved surface, the angles of a triangle always add up to less than 180 degrees, and it is possible to draw a straight line and mark a point not on that line, through which you can draw many more lines, none of which crosses the first line and all of which are, therefore, parallel to it. (140)

Gauss and a now famous student named Bernhard Riemann developed "spherical geometry," which mathematically explains lines on a curved surface. Gribbin explains this through the idea of longitude lines, which are parallel at the bulge of the globe but intersect at the North and South Poles. In "The Masterwork" section, Gribbin describes the importance of "tensors" in Riemann's geometry. Tensors are to Riemann what vectors are to Euclid (the geometrically uninitiated might like to know at this point that vectors to Euclid are measurements of an object's direction and speed). Gribbin writes,

The distances between points in Riemannian space are calculated in terms of a particular kind of tensor, known as the "metric tensor." The metric tensor that applies to curved four-dimensional spacetime has been described as a "vector on steroids" and has sixteen components of one another. (146)

Gribbin explains that the special theory is a theory for Euclidean space, while the general theory applies space-time with a curvature. The general theory contains the special theory in the same way that Riemann's descriptions contain Euclid. The mathematics wore Einstein out, and he quit on it. "This turned out to be a mistake," Gribbin notes. "The equation he rejected in 1912 was very nearly the solution he had been looking for" (148). The theoretical world now owned Einstein almost completely, and he worked in isolation from everyone except for the occasional visits from his wife Elsa. The concept of the "lone genius" is not a myth, but everyone can miss something important. Spinning from an inability to make sense of Mercury's orbit with his theories, Einstein "looked again at the calculations from 1912 [and] saw almost immediately where he had gone wrong, and how with a relatively minor tweak he could come up with properly covariant field equations" (150).

Einstein's Alfred Russel Wallace was named David Hilbert, and when the two corresponded after Einstein saw the pathway to general relativity, Hilbert wrote back that he had taken the same path to a different destination. Einstein almost did not finish. Physics had taken a physical toll; he fell sick and did not visit Hilbert but dragged himself back onto the path, and

> reworking the calculation of the orbit of Mercury with the revised version of this theory gave the right answer! The theory now predicted a perihelion of 43 seconds of arc per century, exactly matching observations. Einstein was so excited that he suffered heart palpitations and had to take a rest. But this wasn't all. The same revision to the theory gave a new prediction for the bending of light by the Sun, not 0.85 seconds of arc as he had previously calculated, but exactly twice as much, 1.7 seconds of arc. (151–52)

This latter result would be proven by Arthur Eddington in 1919. Einstein's theory worked in the inverse by explaining an already known phenomenon, and it worked directly by creating a prediction that could be proven through experiential evidence. As for Hilbert, Gribbin writes, "Was it all Einstein's work? The evidence suggests that it was, although Hilbert was within a hair's breadth of getting there first" (153).

There's more to Gribbin's book in the "Legacy" chapter, which explains what the general theory has been used for since 1915. However, at this point, the book review must end. Gribbin's books about Einstein and the general theory present such a new and important conception of relativity theory that it brings up new questions and possibilities. Late in the book, Gribbin lightly jabs at "amateur theorists who delight in trying to find a better theory than Einstein's" (191). And so caution will be exercised and a caveat stated: the following paragraphs do not represent a possibility for a new theory, but rather a few questions raised from Gribbin's interpretation.

Gribbin has written a definitive work on Einstein, his intellectual process, and the creation of a true masterwork. Even though no one still fully understands the wave–particle duality of light, because of Einstein, physicists do understand light's interconnections with space, time, gravity/acceleration, and speed through the general theory of relativity. Because of John Gribbin, Einstein's process and works are now accessible to a much larger number of readers. In fact, Gribbin writes so well that his book takes on its own paradox. By attempting to shine a light on the genius of Einstein, Gribbin further illuminates his own.

V

Wittgenstein and an Incomplete Circle

WHEN LUDWIG WITTGENSTEIN (1889–1951) WAS ONLY 22 YEARS OLD, he went to Cambridge University and burst in on Bertrand Russell while the latter was trying to enjoy a cup of tea with a colleague (C. K. Ogden for trivia lovers). It is hard for modern people to understand the interactions of the upper class prior to World War I; Wittgenstein hailed from one of the wealthiest families in Austria, and Russell's family lineage was one of the wealthiest and most storied in Britain, but apparently it was okay for noblemen of different ages, nationalities, and sexual orientations (no need to go much deeper into that) to just pop into a tea and start arguing about metaphysics.

Whatever the case, Russell interpreted Wittgenstein's precociousness as genius. Russell might have been flattered by the visit. After all, Wittgenstein believed Russell to be a great thinker and worthy of discussing the deepest thoughts. Throughout his life, Wittgenstein possessed the ability to bewitch people with his mannerisms and the intensity of his conversation. Everyone who met him seemed haunted by his presence, an effect only enhanced by the cryptic nature of his writing. Wittgenstein's self-hating homosexuality might have manifested itself as indifference to the reception of his work; he seemed trapped in an internal world of reproach and epistemological thought.

Wittgenstein fought for Austria in World War I and managed to psychologically tunnel his way out of the trenches by writing the *Tractatus Logico-Philosophicus*. It was hard—is hard—for anyone who can picture Wittgenstein hunkered down with a pen and notebook while a war raged around him not to be swept up in a romantic notion of a young and alienated philosopher, pressured by the ultimate expression of Sturm und Drang, generating a great philosophical truth.

The *Tractatus* was not published until 1921; the author's difficulty in finding a publisher might have had something to do with the quasi-religious format of his work. Wittgenstein seemed to mimic the Koranic sura or the biblical chapter and verse with his format. Or one could read the *Tractatus* as a series of interlinked aphorisms, a Tao for mathematics.

The significant sections of the work are as follows:

2.063 The total reality is the world.

2.1 We make to ourselves pictures of facts.

2.11 The picture presents the facts in logical space, the existence and nonexistence of atomic facts.

2.12 The picture is a model of reality.

2.13 To the objects correspond in the picture the elements of the picture.

2.14 The picture is a fact.

2.15 That the elements of the picture are combined with one another in a definite way, represents that the things are so combined with one another. This connection of the elements of the picture is called a structure, and the possibility of this structure is called the form of representation of the picture.

2.151 The form of representation is the possibility that the things are combined with one another as are the elements of the picture.

2.1211 The picture is linked with reality; it reaches up to it.

2.1512 It is like a scale applied to reality. (2009, 9)

The reference to "atomic facts" is an updated version of Immanuel Kant's "thing in itself," which is to say that an object must be permanent and made of material before it can be perceived by an observer. If the smell of rotten meat can be repulsive to humans but attractive to vultures, we cannot really state that the adjective "disgusting" is endemic to the meat. Rather, the word "disgusting" is a reference to the feeling that occurs between the gas and the human. The same interaction between the gas and the vulture would be deemed "attractive."

Wittgenstein essentially states an axiom: that there are atomic facts before human perception but that human perception creates a mental model that fits, without gaps, on those facts. Wittgenstein's philosophy is a direct ancestor to the modern theoretical concept of model-dependent realism, which synthesizes neuroscience with physics to state, essentially, that the human mind takes in sensory data from the universe and then turns those data into a model that helps us to survive and mate. Our mathematical and physical models (theories) of the universe are extrapolations of those initial biologically based images.

Probably no other philosopher excites debate like Wittgenstein. In his lifetime, he was considered both a genius and a fraud, and so it still is. The *Tractatus* is not fraudulent, nor is it pseudo-philosophy but rather an important attempt to connect mathematical axioms to the physical realities, to join the Platonic realm with the real world. He didn't solve the problem of the axiom—he merely moved it to the senses. His arguments hold only as long as the mathematical models seemed to measure up, without gaps, to the reality.

Even as Wittgenstein's words were published, the commutative aneurysm that got tangled into Western mathematics by Cardano in 1545 throbbed ever more urgently. It would burst onto a desolate island in the North Sea called Heligoland, when Werner Heisenberg adapted matrix algebra to quantum physics, but that story will be told in conjunction with the development of quantum physics and the creation of the incompleteness theorem.

In 1924, a German physicist, philosopher, and professor named Moritz Schlick (1882–1936) organized a group of philosophers and mathematicians as a way of picking up on the work of what had been called, before the war, the Ernst Mach Society. (The name "Vienna Circle" did not appear until 1928, but it will be used consistently here for clarity's sake.) The Vienna Circle included the mathematician Kurt Gödel (1906–1978), who absorbed the group's "logical positivism." Essentially, "logical positivism" sought to bring an engineer's clarity and practicality to philosophy. Epistemological attempts to justify the logic of mathematics were treated as pointless abstractions, akin to theology.

The Vienna Circle read and discussed Wittgenstein's *Tractatus* but seemed interested (as has almost everyone who has discussed that work since) primarily in that work's cryptic final phrase, where Wittgenstein states that if there can be no knowledge of something, then one should remain silent. It was an unfortunate ending to an original work because it created an easy philosophical out for anyone not wanting to engage with the problems of axiomatic and infinite mathematics.

For a philosophical book club that met in a coffeehouse, the Vienna Circle accomplished a lot. Their contributions range from historical analysis to the development of a new school of philosophy to world-altering sociological studies. As is often the case with philosophy, drama between the philosophers (such as a notorious incident in 1946, when Ludwig Wittgenstein shook a fireplace poker at Karl von Popper) and tragedy (a crazed philosophy student murdered Schlick in 1936) helped to memorialize and sensationalize the group. Readers who come for the action might stay a little while for the philosophy.

At this point, the histories of physics and mathematics start to merge into a muddle. As has already been stated, Max Planck established in 1900 that radiation could best be mathematically understood if one assumes that light is emitted through packets of radiation rather than in waves. In 1913, Niels Bohr expanded on that insight by applying it to the atomic structure.

Prior to 1913, physicists and chemists had gotten comfortable with Ernest Rutherford's (1871–1937) "planetary" model of the atom, which showed electrons orbiting a nucleus. The solar system analogy proved to be a little too handy, though, if one considered that internal energy, not gravity, keeps the electrons in their orbits. Bohr, to quote George Gamow in his *Thirty Years That Shook Physics* (1985),

> calculated the energies of various discrete quantum states of atomic electrons and interpreted the emission of light as the ejection of a light quantum with energy equal to the energy difference between the initial and final quantum states of an atomic electron. With his calculations he was able to explain in great detail the spectral lines of hydrogen and heavier elements, a problem which for decades had mystified the spectroscopists. Bohr's first paper on the quantum theory of the atom led to cataclysmic developments. Within a decade, due to the joint efforts of theoretical as well as experimental physicists of many lands, the optical, magnetic, and chemical properties of various atoms were understood in great detail. But as the years ran by, it became clearer and clearer that, successful as Bohr's theory was, it was still not a final theory since it could not explain some things that were known about atoms. For example, it failed completely to describe the transition process of an electron from one quantum state to another, and there was no way of calculating the intensities of various lines in optical spectra. (2)

The transition process of electrons from one stage to another eventually results in the release of energy, and according to Planck, this happened in quanta. There is, however, no way of putting the quanta back where it was, meaning that quantum states do not commute. In 1925, while contemplating this at Heligoland, Werner Heisenberg applied matrix mathematics, a technique he learned from Max Born, as a way of describing the forward-only motion of quantum interactions. All chemical and atomic interactions result in a loss of heat. This process can be sped up or slowed down, but not reversed.

Now, matrix algebra seems like just another chapter in dry high school textbooks. Students often see it as just another arbitrary set of mathematical rules to be memorized and applied for a test. Teachers and professors frequently skip matrix algebra because it seems to distract from the teaching and understanding of more traditional mathematical frameworks, but matrix algebra forced mathematicians into a reckoning: mathematics cannot tell nature what to do; to be effective, mathematics must be modeled after what nature does. Mathematics is human-made, not intuited from a Platonic realm of logic.

And if that is true, then mathematics contains axioms that cannot be proven. Gödel was poised to write his famous paper.

Gödel's famous incompleteness theorem might best be seen as a footnote to Wittgenstein. It's a logical extension, complete with mathematical symbols, of how there are certain things that mathematics must be silent about. To be clear, there is no evidence that Heisenberg's matrix algebra influenced Gödel's thoughts; instead, the sick and depressed Gödel seems to have become immersed in the Vienna Circle's fascination with the axiom and infinity. The fact that his 1931 paper so closely followed Heisenberg's 1925 quantum application of matrix algebra might just be another of scientific history's examples of a sudden convergence of ideas in the heads of different thinkers at the same time, such as was the case with Leibniz/Newton and the calculus or Darwin/Alfred Russel Wallace and evolutionary theory.

Gödel begins his paper by directly addressing *Principia Mathematica* and the problem of the axiom:

> The development of mathematics towards greater exactness has, as is well-known, led to formalization of large areas of it such that you can carry out proofs by following a few mechanical rules. The most comprehensive current formal systems are the system of Principia Mathematica (PM) on the one hand, the Zermelo-Fraenkelian axiom-system of set theory on the other hand. These two systems are so far developed that you can formalize in them all proof methods that are currently in use in mathematics, i.e., you can reduce these proof methods to a few axioms and deduction rules. Therefore, the conclusion seems plausible that these deduction rules are sufficient to decide all mathematical questions expressible in those systems. We will show that this is not true, but that there are even relatively easy problems in the theory of ordinary whole numbers that cannot be decided from the axioms. This is not due to the nature of these systems, but it is true for a very wide class of formal systems, which in particular includes all those that you get by adding a

finite number of axioms to the above mentioned systems, provided the additional axioms don't make false theorems provable. (2)

Gödel's paper is just a few pages long but dense with symbology. One wonders how Bertrand Russell and Alfred North Whitehead felt about seeing such a short work as a counter to the multivolume study they spent years of their lives working on, but Gödel effectively made the case that mathematicians cannot use mathematics to prove mathematics. If that is true, then mathematics must be a secondary function to reality. In 1925, Heisenberg studied the noncommutative nature of quantum heat loss and created matrix algebra to describe quantum interactions, and in 1931, Gödel created a proof stating that it was okay to do that.

The incompleteness theorem unchained physics from Platonic mathematics and allowed the physicists to stop staring at shadows on the wall and turn around to look out at the real world. Gödel's paper also created a climate where computer programmers recognized that they could develop symbols that gave computational instructions to a computer. Much has been made about how dense Gödel's writing was, but it could be that only a deeply depressed and outwardly strange mathematician could write in such a form that it would attract the intellect of someone with similar characteristics. It takes little imagination to understand why Alan Turing (1912–1954) would find a direct analogy between his work with the Colossus computer at Bletchley Park and Gödel's theorem.

In his book *The Modern Mind: An Intellectual History of the 20th Century* (2001), Peter Watson shows the connection:

> Turing conceived what he called a universal machine, a device that could perform all calculations. Finally (and this is where the echo of Godel is most strong), he added the following idea: assume that the universal machine responds to a list of integers corresponding to certain types of calculation. For example, 1 might mean "finding factors," 2 might mean "finding square roots," 3 might mean "following the rules of chess," and so on. What would happen, Turing now asked, if the universal machine was fed a number that corresponded to itself? His point was that such a machine could not exist even in theory, and therefore,

he implied a calculation of that type was simply not computable. There were/are no rules to explain how you can prove, or disprove, something in mathematics, using mathematics itself. (363)

(An alert reader will also likely see the connection between Gödel's theorem and the problems with infinity stated earlier.) Gödel operates historically as the philosophical antithesis to Plato and the founder of a process that led to the development of computers and thermonuclear weapons.

Essentially, Turing found that you cannot code a machine to tell you when it becomes sentient. The only way that the machine could judge that is by its own internal standards and definitions. It could declare itself to be sentient anytime without any actual thought or emotion. By the same line of logic, you cannot ask mathematics to tell you when it is true because mathematics can judge such a statement only by its own internal lines of mathematical logic.

W

Planck, Bohr, the Curies, Heisenberg, and Uranium

BEFORE EMBARKING ON THIS SECTION, IT SEEMS PRUDENT TO REVISIT the thesis of this book. The goal here is to untangle the historical narrative to put "discoveries" in an order that makes sense of understanding the totality of science. The point is, emphatically, not to say that science cannot advance despite a confusing narrative.

For example, the term "energy" causes confusion but not so much confusion that it delayed the creation of atomic weapons. The history of science shows that models cannot be "wrong," just more or less descriptive. For practical purposes, if one sees the splitting of a uranium atom as releasing entropy or energy, the difference would be small. It's only when trying to reconcile that process with commutative algebra or stoichiometry that a problem arises in trying to develop a unified understanding.

The entire concept of radiation, developed by the Curies, indicated that the traditional atomic model, indeed, the classification methods used to arrange elements on the periodic table, indicated a static picture of the atomic realm. Something happened in the early 20th century that was similar to the development of Darwinian evolution in 1859 and Clausius's expression of the second law of thermodynamics in 1860. That is, about the same time (1904) that J. J. Thompson synthesized his 1897 finding of the electron into a model of the atom, this model, either derisively or playfully, came to be called the "plum pudding" model, but it was based on a solid mathematical conceit: that an electron is negatively charged but randomly skitters about in a positively charged mass of the atom.

In these years, the physics world attempted to absorb and understand the time-warping concepts at the core of special relativity. In addition, it might need to be said that it became fashionable for mathematicians and scientists to develop a reputation from their ability to express an understanding of the strangeness inherent in relativity theory. If the E in Einstein's equation had been understood as entropy, then Ernest Rutherford's (1871–1937) 1911 paper "The Scattering of Alpha and Beta Particles by Matter and the Structure of the Atom" would have been viewed as definitive proof for Einstein's assertion.

As it was, Arthur Eddington's 1919 observation of a solar eclipse provided evidence for Einstein's theory of gravity. The fascination with gravity, of course, relates to Einstein's historical relationship with Newton. For about two and a half centuries, physics was associated with gravitation studies, so, of course, physicists showed the most interest in the gravitational components of relativity theory.

Because Aristotle focused on motion, Newton asked why the apple fell rather than why it rotted, and because of that, Einstein did as well, and Eddington (and the public) focused on the gravitational component of relativity.

If Newton had asked why the apple rotted, then Clausius would have been Newton's successor and Einstein's insight seen as being related to entropy. In that case, Rutherford would have been viewed as giving evidentiary proof to relativity.

Rutherford's famous experiment involved firing alpha radiation at a sheet of thin (*very* thin, like 1,000 atoms) gold foil. Surrounding the gold foil was a screen made of zinc sulfide that could catch the alpha particles as they either went through the gold foil or bounced back from it. The alpha particles mostly passed through the foil, but there was a 1 in 8,000 chance, calculated afterward, that the alpha radiation would bounce back from the gold foil. These results indicated that an atom contains mostly empty space but held a positive charge somewhere in the core that contained mass: a nucleus.

What followed is the well-known evolution of the atomic model. Rutherford's 1911 model envisioned electrons swirling around a dense nucleus, forming orbits of interconnected rings that still are used in popular representations about nuclear physics. It was not until 1913 that Niels Bohr arranged electrons into a series of "shells" around the nucleus in a model that featured only the nucleus and the electrons. It was not until 1932 that the English physicist James Chadwick (1891–1974) published two back-to-back papers, the titles of which express the progress of his research. The "Possible Existence of Neutrons" was followed up by "The Existence of a Neutron."

Although Chadwick was a student of Rutherford's, he based his research on the work of Irene Joliot-Curie (1897–1956) and her husband Frederic (1900–1958). Irene Joliot-Curie provides further proof for the occasional genetic inheritance of genius, as her famous parents had initially discovered atomic radiation. Marie and Pierre Curie discovered that certain elements create their own radiation, from their core, at a level that can be detected by human instruments. (Elements rot like apples but only on a different timescale.) Irene Joliot-Curie and her husband discovered, to quote Peter Watson (2001), that "by bombarding medium-weight elements with alpha particles from polonium, they had found a way of making matter artificially radioactive" (393).

But this was only part of the story. For "light" particles, outside radiation knocks protons and/or alpha particles out of orbit, causing them to emit heat and become lighter. What happened to heavier particles under the same bombardment was not discovered until Enrico Fermi understood beta decay.

This means that Joliot-Curie and her husband had discovered how to manipulate entropy at the atomic level. This won them a Nobel Prize at the time but might be seen as one of the most significant discoveries in the history of science, the true moment that humanity stole fire from those clever gods who thought to disperse and hide it in the smallest of places.

Practically speaking, "artificial radiation" allowed for the transmutation of elements. If an element is defined by the number of electrons in the orbital shells, then forcing those electrons into emission changes the element. (Bananas go from green to yellow to brown to black. You can slow that down by freezing the banana or speed it up by burning it.) At the atomic level of explanation, the neutron interacts with the nucleus, releasing radiation, which can also be seen as decay. From a thermodynamic perspective, the neutron forces an interaction, and all interactions lead to the loss of heat, and heat is the equivalent of radiation.

But what kind of radiation? Chadwick did not accept the assumption that the decay occurred as a gamma ray. Instead, he developed hypersensitive detection devices and measured the radiation at a more precise level than Joliot-Curie and her husband could have. His measurements indicated that the emitted radiation was equivalent in mass to the protons. This meant the radiation was made up not of gamma rays but of neutrally charged particles in the nucleus that he called neutrons. This led to the creation of the 1932 model of the atom that included protons and neutrons intermingled in the nucleus, with electrons orbiting in Bohr's shells.

This understanding allowed physicists and chemists to see a new species of element known as the isotope. An element such as uranium is marbled with atoms of different types, defined only by the differences in the number of neutrons. Elements such as hydrogen could be altered in different types of isotopes, such as tritium or deuterium, depending on the number of neutrons.

In 1933, the Italian physicist Enrico Fermi (1901–1954), working out of the University of Chicago, first introduced the idea of "weak interaction." The "weak interaction" manifested itself as beta decay and this eventually defined two forces at the subatomic level, simply called the weak force and the strong force. At the macro level, gravity and electromagnetism defined the interactions between the larger bodies (sizes ranging from a germ to Jupiter are all the same when compared to the tininess of the atomic level). (The search to unify these separate forces would take on greater historical importance later in the 20th century as physicists threw themselves into black holes and found out that, if nothing else, they were filled with money.)

Fermi also found that heavier elements acted differently than lighter elements when bombarded with outside neutrons. The bonds in the nucleus of heavier elements absorbed incoming neutrons, thus gaining mass. The extra mass, however, created a new isotope not made anywhere by nature.

And since an element not seen anywhere in nature will still be subject to the second law of thermodynamics, it will decay. But it will decay in a way not seen before.

The German physicist Otto Hahn (1879–1968) proved that bombarding heavy elements with neutrons caused the nucleus of the element to absorb the neutron. If uranium-238 had a core nucleus that broke apart when hit by neutrons, it would dissolve into radium, or radium-230. However, when neutrons were shot into uranium-238, the result was always the lighter element of barium. Hahn expressed his puzzlement to Lise Meitner (1878–1968). They eventually decided to apply a model of Bohr's. As Watson (2001) writes,

> They began to consider . . . in particular a theory of Bohr's that the nucleus of an atom was like a drop of water, which is held together by the attraction that the molecules have for each other, just as the nucleus is held together by the force of its constituents. Until then . . . physicists had considered that when the nucleus was bombarded, it was so stable that at most the odd particle could be chipped off. (395)

With the "drop of water" model, the attractions between molecules could be cut from each other, or split, in a way that would cause the uranium to decay into barium and krypton. Watson (2001) writes of the crucial finding,

> As news sank in around the world, people realized that, as the nucleus split apart, it released energy, as heat. If that energy was in the form of neutrons, and in sufficient quantity, then a chain reaction, and a bomb, might indeed be possible. Possible, but not easy. Uranium is very stable, with a half-life of 4.5 billion years. . . . It was Bohr who grasped the essential truth—that U238, the common isotope, was stable but U235, the much less common form, was susceptible to nuclear fission. (395–96)

A nuclear bomb was suddenly theoretically possible, but it was really an entropy bomb, and no one knew how much uranium would be needed to build it. The story of the Manhattan Project and the movement from atomic theory to an atomic explosion is well known enough to not need to be repeated here. However, atomic explosions might be seen as extrapolations of thermodynamic processes at the core of the atom. In the same way that the 1969 moon landing was based on equations established almost three centuries earlier by Newton, the atomic bomb, with the possible exclusion of von Neumann's compression concept, had its theoretical foundations established by 1933.

X

Uncertainty and Exclusion

By 1933, physicists and chemists understood the two macro-level forces of electromagnetism and gravity and the two micro-level forces of the strong and weak forces. They understood that protons and neutrons formed the core of the atomic nucleus, that the electrons surrounding the core resided in shells, and that electrons "leapt" between the shells. When energy escaped the shells, it did so in the form of alpha, beta, or gamma rays, which practically manifested themselves as heat. They knew how to shake the core of an atom with neutrons either by chipping away at light elements or by forcing heavier elements to become heavier through neutron absorption. All of this made for a practical and workable physics, but Einstein had reminded everyone that physics is not only about the observed object but also about the observer.

Questions about how the observer can "know what she knows" would eventually lead quantum physics to rely on probability and statistics, but that break came as a result of Heisenberg's uncertainty principle and Paul's exclusion principle, both of which provided limits to human knowledge at the atomic level.

In 1925, Wolfgang Pauli (1900–1958), an Austrian physicist who seemed to be a lovable character, discovered the exclusion principle. For the most part, this narrative has tried to avoid biographical sketches of the main thinkers, but in his classic *Thirty Years That Shook Physics: The Story of Quantum Theory* (1985), the Russian physicist George Gamow (1904–1968) wrote an irresistible description of his friend that is worth the aside:

Once, presumably on a doctor's order, Pauli decided to lose weight and, as in everything else he attempted, succeeded very quickly. When, minus a number of pounds, he appeared again in Copenhagen, he was quite a different man: sad, humorless, and grunting instead of laughing. We all urged him to join us in delicious wiener schnitzel and good Carlsburg beer, and in less than a fortnight Pauli became again his old gay self. (63)

Gamow goes on to relate a story about Pauli's anti-Nazism. Pauli broke his arm while coming out of a boat, and the cast made his arm stand straight out. Pauli refused to get his picture taken with the cast on because he thought it looked like he was doing the "Sieg Heil."

Two years after his epiphany about matrix algebra at Heligoland, Werner Heisenberg established the uncertainty principle. This is an important concept later connected to the "spooky action at a distance" notion, which later becomes such a staple in works of popular physics. Pauli wanted to know what factors determined the energy level of an electron and also wanted to discover the number of electrons that each of Bohr's shells could hold.

Pauli found that electrons carry information about themselves that manifests itself in four separate categories. An electron defines itself by its spin, energy, angular momentum, and magnetism. If two electrons are in the same shell, they cannot share the same characteristics. If two electrons reside in the same shell, then they must have opposite spins. This means that if you could measure the spin of one electron, then you automatically would know the spin of the other electron. By analogy, if you are facing north and sitting at a red light, then you know that the light for vehicles traveling east to west or west to east will not be red.

Even if two electrons from a shell are separated, they will always have opposite characteristics, so when you know the spin of an electron under a microscope, you can gain insight into the spin of an electron thousands of light-years away. The information connection between the two is not confined by distance.

About the same time, another Austrian physicist, Erwin Schrödinger (1887–1961), began to write about the emission of quantum particles as existing in waves of probability. According to this theory, the law of large numbers is what gives the "real" world its permanence. In the aggregate, atoms in stable elements tend to act pretty much the same from day to day. However, when an electron is radiated outward as an alpha, beta, or gamma ray, it has to be treated mathematically as a singular particle, which means that the law of large numbers no longer applies. The emission can be anywhere, and its position can be predicted only via probabilities.

British physicist Paul Dirac (1902–1984) took Pauli's exclusion principle to a logical (if uncomfortable) extreme in 1931 by stating that if electrons had to manifest different characteristics in the same orbital shell, then the same must be true for all matter, which indicated that matter must be balanced by the existence of antimatter. He was also the first to posit that $E = mc^2$ could be explained through the lens of thermodynamics. It did no violence to Einstein's famous equation if the E stood for entropy.

In 1935, Schrödinger constructed a macabre Rube Goldberg–ish thought experiment involving a cat that would become the most famous in scientific history, even more so than Maxwell's exciting demon. Schrödinger imagined a cat put into a box that contained a cyanide capsule, a radioactive atom, and a Geiger counter. The atom radiates on average once every two hours, which means there is a 50 percent chance that it radiates during the one hour that the cat must endure this experiment. If the atom radiates, the particle trips the Geiger counter, and then the cyanide capsule bursts and kills the cat.

The purpose of the experiment is to extrapolate the atomic uncertainty, a micro event, to a macro event. The entire chain of events therefore gets defined under the same probability wave as the radiated particle. The cat, therefore, is subject to the same "superposition" of two possible states that the atomic radiation is subject to. The cat can be either dead or alive, just like an electron can spin either up or down. The discovery of a spin-up particle collapses the probabilities for the other particle and ensures that it will be a spin-down particle. Local information gives you sudden clarity about distant information.

Until you open the box, the cat exists in a superposition of being either dead or alive. When you discover the kitty's fate, the moment of your discovery retroactively collapses the probability function into the cat being either dead or alive. If the cat meows at you, then the probability wave collapses to the atom holding on to its radiation. If the cat is inactive, it's because the probability wave collapses backward and allows for the emission of radiation.

The thought experiment is misleading for two reasons:

A. The experiment is time dependent. Leave the cat in the box for long enough, and the atom, any atom, will inevitably emit a particle. Time is a relative concept that requires a reference point. Entropy does not require a reference point: there is either more or less of it. Even if you open the box and save the cat, the cat itself will radiate away before long, and nine lives won't stave off entropy.

B. The experiment ignores that probability functions exist, like time, not as an actual thing but as a relationship between the observer and the observed. To use an explanatory example, we might say "the couples dance" as a noun, but a "couples dance" is actually a term used for a relational activity between two people; it is not a thing in and of itself. You cannot have a "couples dance" with one person. You cannot have a probability wave function without two objects interacting.

In logic, treating a concept, or a relational conceit, as a noun or "thing" is an error called reification. The problem is not confined to abstractions like probability theory; it shows up in more concrete calculations. For example (and this will be important later), gravity is not a thing. Gravity is a relational concept. The gravitational equations plainly states this, but no one ever sees it. The gravitational equation states that "mass 1" multiplied by "mass 2" and divided by the radius between them squared equals the force of gravity. But you can't have gravity by this definition if there is just one object. By gravity's own definition, gravity cannot exist in a singularity.

What Heisenberg determined, eventually, was that in a relational concept, the observer would always be forced to confront a certain amount of uncertainty. His principle is often misstated, and like a lot of conceits about quantum physics, the core notion is made to seem more mysterious than it is. Schrödinger merely stated that it is not possible to know with certainty both the position and the momentum of a particle. The more you know about the momentum, the less you know about the position.

Heisenberg's principle says nothing that a statistician couldn't say in terms of a "confidence interval." Momentum, by definition, lacks a direction; if it had a direction, it would be defined as velocity. If an observer tracks a particle's momentum, then the observer becomes less and less confident (but not clueless) about the absolute position of the particle. The issue becomes a seesaw of confidence levels, not an absolute statement about total uncertainty. But this is true of most things in statistics. The more general a prediction is, the more likely it is to be true. The more specific the prediction, the less likely it is to be true.

The uncertainty principle is another one of these scientific concepts that suffers from an exciting name. In boring physics terminology, a particle is smaller than the wavelength of the light that bounces off it. To "see" a quantum particle, one must bounce gamma radiation off it. Gamma radiation comes in short but intense frequencies that are strong enough to knock a particle off its course. Light is weak enough to leave a quantum particle in place but too big to find the particle. Gamma radiation is small enough to find the particle but too strong to leave the particle's position unaltered.

Unlike Pauli, Heisenberg seemed to exhibit no overt hatred for the Nazis. In fact, Heisenberg led Nazi Germany's atomic weapons program. Fear of Heisenberg's intellect is what drove Einstein and other notable physicists to lobby President Franklin D. Roosevelt to start an American atomic bomb project. Heisenberg managed, somehow, to exist in a dual state throughout World War II.

On the one hand, he led the Nazi effort to create an atomic bomb and spent those dark and dangerous years as a respected scientific administrator. On the other, he did so little in that role that he could later plausibly claim to have been secretly anti-Nazi all along. By refusing to assert his immense intellect to develop a nuclear weapon for Hitler,

Heisenberg could claim to be a great hero for humanity. His ability to exist in an uncertain and dual state, where he could claim credit as a great man regardless of the outcome, demonstrates that his cunning matched his theoretical genius. He could not claim, however, to have an ethical sense that matched his theoretical skill.

All of this confusion led to the chief error in quantum physics: the concept of wave–particle duality, a concept derived from the double-slit experiment. The double-slit experiment has a long history but was demonstrated in its modern quantum context in 1927 at Western Electric (later known as Bell) labs. The experiment involves a concentrated light source and a screen. A plate with two side-by-side slits is placed in between the light source and the screen. When the light source is turned on, the wavelike properties of light go through the slits and interfere with each other, creating bars of light and shadow on the screen. This would seem to prove that light is a wave.

However, the screen detects the light as push points, as if light came in packets or quanta that had some mass as particles. If light were a wave, it would pass through both slits at once. If it were a particle, it would need to pick a slit. If an observer can detect which slit the particle passes through, then the light beam will not form an interference pattern. If you watch light, apparently, it acts like a particle. If you don't, then it acts like a wave.

No experiment in scientific history has been misused more by mystics than the double-slit experiment. People who half understand it jump to the conclusion that the universe exists in the form of probability only until it is observed. Some desperate theologians even claim that the double-slit experiment (finally) produces a logical proof for the existence of God since an all-powerful observer would be necessary to break down probability waves.

Yet quantum particles are not doing anything mysterious. If you peer down to a small enough level, you will find the limits of your powers of detection. Light does not act like either a particle or a wave; it acts like light. Physicists had to describe this new property via analogies that had been established via already known phenomena. Once an observer's ability to detect something scientifically is exhausted, the observer must shift to probabilities (reified and dressed up with statistics, these become strings).

The problem with looking down at something as fast and small as quantum particles/waves is that they move a lot faster than you do. Observers are never looking at where light is, only at where it has been. It takes the average person about three seconds to analyze and make sense of sensory information. You can observe light only where it was three seconds ago, not where it is now. The atomic facts of the universe are always uncertain to observers. It is not possible to think about those atomic facts in a way that is not relational between the observer and the observed.

Quantum physics, as a discipline, developed before neuroscience and studies of the mind matured, which is why the effect of the observer on quantum particles has not been fully understood. For example, it might seem to defy logic that an electron can leap from one state to another in an orbital shell without going through an intermediary stage. This is because humans think that if a plant grows from one to two inches, then it must go through a stage where it was one and a half inches. The reality is that such an observation is contained not in logic but in common sense. Our senses evolved to help us survive, not to understand quantum phenomena, so our brain fills in gaps to make the picture of atomic facts sensible.

This will be explored further in the chapter "The Third Law of Thermodynamics, String Theory, and Model-Dependent Realism" in a discussion about model-dependent realism, but think about taking a picture with an old Polaroid camera. Whether the picture is blurry or clear depends on how you held the camera at the time the photograph was taken, but by the time the picture develops, the subject of your picture is somewhere else. When you are looking at the picture, you are not looking at anything as it currently exists. The problem *is in the picture*, and if you don't take the picture, the problem does not exist.

Y

Convergence:
Relativity and Thermodynamics

IT'S PROBABLY BLASPHEMOUS (OR SOMETHING CLOSE TO IT) TO SAY that relativity theory created an unnecessary diversion in the progress of scientific understanding. Alternative histories are pointless, but Einstein developed thought experiments and equations based on relational understandings that would end up causing havoc on future theoretical models.

The concept of time dilation, for example, is mind-bending but not that complex. A clock is just a ruler bent into a circle. An observer's "experience" of a mile is determined by how fast the observer is going. Time does not measure entropy; temperature does. If we used temperature to determine the rate of an object's decay, we could then skip the concept of time and just see objects in relation to their overall level of decay.

According to time dilation, if we have two twins (we'll call them Cain and Abel for fun), we can split them up at different speeds, and they will age at different rates. If Cain gets on a spaceship and gets near the speed of light, this will elongate his frame of reference. If Abel stays on Earth, his frame of reference will be the same as experienced by all other terrestrials.

Imagine two large boxes (frames) on top of 730 smaller boxes, both of which are the same total length. The boxes represent days. At nearly the speed of light, Cain flies through two boxes and then comes down in the 730th box, where he will encounter his brother Abel, who went through 730 boxes. Cain will be two days older and Abel two years older. How is this possible?

We can understand the boxes as being defined by their rate of entropy. Cain radiated only two days' worth of heat, while Abel radiated 730 days' worth. The same effect could have been achieved by reducing the temperature of Cain and his spaceship to nearly 0 Kelvin (K). As would be the case with the speed of light, the closer Cain and his spaceship got to 0 K, the less entropy they would experience.

A lot of brain power was, therefore, used up (wasted?) in trying to equate the more direct principles of thermodynamics in terms of relativity. The big problem is that because time is a relational conceit, the reversal of time means the reversal of motion, and that wreaks havoc on theoretical models at the boundary conditions.

As early as 1937, physicist Richard C. Tolman (1881–1948) of Caltech wrote a book titled *Relativity, Thermodynamics and Cosmology* (2011), where he explained how thermodynamics was compatible with special and general relativity but, in two separate passages, gently indicated that thermodynamics provided a better cosmological model because it could not be reversed. These are two long and dense paragraphs, but Tolman deserves credit for his insights. These two sections might at first seem dense and unrelated, but because they deal with inflationary properties, they are indispensable:

> Hence the application of the relativistic second law of thermodynamics to such cosmological models shows that the proper entropy of each element of fluid as measured by a local observer can in any case only increase or remain constant. Since the equality sign applies to reversible processes, constant proper entropy for each element of fluid then becomes the necessary requirement for the reversible expansion of those models at a finite rate.
>
> This requirement, however, can apparently be met provided the fluid filling the model is taken as sufficiently simple.
>
> To show this, we first note that the nature of the model is such as to eliminate possibilities for entropy increase which might otherwise result from the interaction between the elements of fluid and their surroundings. Thus there is no increase in entropy due to the irreversible

heat flow throughout the model; there is no entropy increase due to the friction of moving pistons or the like since no containers for the elements of fluid are now involved; and there is no entropy increase due to an inability of the fluid to keep up its full pressure at the expanding boundary of an element owing to the uniform pressure throughout the whole model. There hence appears to be no irreversibility directly due to poor coupling of any element of fluid with its surroundings.

The remaining possibilities for entropy increase then lie in irreversible changes taking place inside the elements of fluid in the actual material of which the fluid is composed. . . . In the case of simple enough fluids, however, such possibilities for internal entropy increase would be almost or completely lacking. Thus if we took a perfect monatomic gas, as suggested for the fluid at the beginning of the section, there would be no possibility for internal irreversible processes, provided we neglect the small energy transfers that would take place between the gas and the slight amount of thermal radiation that would actually be present. . . .

The relativistic discovery of cosmological models, filled with material throughout their entire extent, thus provides possibilities for the expansion of an unenclosed fluid without its dissipation into empty space and without friction, irreversible heat flow, or pressure drop at the walls of any container, of a kind hitherto unknown. Analyzed from the point of view of relativistic thermodynamics, this then leads to an increased possibility for processes to take place at a finite rate and yet also either with complete reversibility, or in any case with the elimination of sources of irreversibility that seemed classically inevitable.

. . . The thermodynamic condition for reversibility, which we have obtained by taking the equality sign in the relativistic expression for the second law, actually agrees with the requirements for a real reversal in the motions of cosmological models; and it will be shown . . . that an observer in a reversibly expanding universe would be led to quite erroneous conclusions if he should try to interpret the behavior of his surroundings by the use of classical rather than relativistic thermodynamics. (324–26)

To translate, a traditional cosmological model based on movement, gravity, and time would indicate an expanding universe. From our current perspective, we could imagine moving backward to a previous position.

If a ball rolls forward, it can be rolled backward. Reversing motion is misleading, however, because the ball that moves back to its previous position is not the same ball. Not only did it radiate, but the interaction of the ball and the surface it rolled on created quantum interactions that released heat in the form of friction. Also, given that everything else in the universe also radiated while you were moving the ball, nothing really got reversed from the perspective of thermodynamics.

By not taking this into account, problems with cosmological models of expansion arise. Expansion indicates a reversibility that does not exist. Tolman writes that

> by considering the integration of the . . . expression for the rate of expansion, it will be seen that any finite value of R would then be reached in a finite time, but that an infinite time would elapse during the passage from R finite to R infinite.
>
> Such a model, which expands without reversal from a singular state in the past to infinity in the future, we designate as a monotonic universe of the first kind, type M1. As a model for the actual universe, it has the disadvantage of spending only an infinitesimal fraction of its total existence in a condition which differs appreciably from that of a completely empty . . . universe. Hence, if we are willing to take our observations on the actual universe as giving a fair sample of the kind of conditions that would be found anywhere and at any time, we should then rule out this model. (399–400)

This was published 12 years before the term "Big Bang" popularized the conclusion created by reversing standard cosmological models based on movement. To "reverse" a cosmological process is to engage in an interaction that would cause heat loss, or entropy. It cannot be done.

It would take a while, but eventually theoretical physicists would find that reversal is impossible within the theoretical framework of relativity theory. If one posits the reversal of all matter back to a dense particle, then equations indicate extreme heat. Heat is always an indicator of movement, and movement is always an indicator of time.

The concept of time requires a steady reference point, and this means that in order to assign a relational concept of time to the movements of the early universe, one needs to posit a Platonic Earth orbiting a Platonic sun in order to develop the reference point of a "year." The universe does not allow for this.

The staircase out of the singularity is exponential. The staircase back to it is logarithmic. If you carry a clock up to the top of the exponential staircase, you can't leave it there and then walk down the logarithmic staircase while looking back to see what time it is. The conditions of gravity work on the clock, and gravity dilates as your radius increases. You have to take the clock with you, in which case you will not experience a change in time any more than you currently experience the daily decrease in gravitational force. (Yes, this is true. If the universe is expanding, then the radius between masses expands as well, diluting the force of gravity, but there is no reference point.)

Imagine you are standing on top of the exponential staircase, and then you tie a clock and a thermometer to a string and toss it down the logarithmic staircase. When you pull the clock and the thermometer back up to where you are standing, they will just show you the time and temperature where you are, not where they have been. This means that all you can do is grab a pen, paper, and a calculator and start working on probabilities.

Everyone knows that Einstein had trouble adapting to quantum mechanics—all that "God doesn't play dice" business—but it could be that he threw off a generation of theoretical physicists who operated in a world of trains, clocks, and twins that did not radiate.

Z

The Third Law of Thermodynamics, String Theory, and Model-Dependent Realism

THE THIRD LAW OF THERMODYNAMICS DEVELOPED GRADUALLY BETWEEN 1906 and 1912 by the German chemist~µ Walther Nernst (1864–1941). In 1905, Einstein's miracle year, Nernst discovered that entropy decreased along with temperature. Entropy can approach 0 K (aka absolute zero) but never quite get there because some level of free energy always stays above absolute zero. If we think of the free energy as latent entropy, or a probability field of electron discharge (aka radiation), then we can see that Nernst laid the foundations for a cosmological model based entirely on three laws of thermodynamics.

The question should not be "Why is there something rather than nothing?" Rather, it should be "Why is the universe not 0 K?" This is because reversing entropy rather than motion brings the equations to a zero-entropy state, which is frozen and surrounded by a probability field for eventual heat radiation. The probability field does not actually exist unless the heat radiation creates conditions for sentient beings who can understand the concept in such a way.

Had Einstein just stamped patents, one could easily imagine that the same laws of relativity would have developed around the laws of thermodynamics. Time would have been understood as a relative measurement of the movements underlying heat loss, with temperature being the direct measure of entropy. The colder you make an object, the more you slow its aging, and you can approach absolute zero only theoretically in the same way that you can approach the speed of light.

The potential for an alternative history of science based around thermodynamic laws was always there but was missed because of the difficulty in rewriting history and nomenclature, although, as will be seen shortly, the scientists who pioneered quantum theory saw the opportunity.

After the creation of fusion and fission bombs, both of which operate by accelerating entropy (aka releasing energy), theoretical physics took a turn toward ever-greater levels of mathematical abstraction. Physicists posited and sometimes found evidence for an increasing number of particles and antiparticles and in the really quantum realm began to discard the idea of particles altogether and focus on the vibrating packets of energy that came to be known as "strings."

In practice, string theory and other theories came to be seen as largely theoretical models, but mathematically speaking, they seem to be a relatively rational response to the quantum. If the macro universe is held together by the law of large numbers (probabilities for quantum stability), the only way to apply that law to a single radiated particle is to split the particle into theoretical vibrating strings that will then act with some probabilistic consistency.

For anyone looking to resist entropy, there was always the possibility that gravity, considered as a force, could prevent elemental decay. When the Princeton physicist John Wheeler (1911–2008) decided to discard the phrase "completely collapsed objects" in favor of "black holes," the public got interested. When Stephen Hawking (1942–2018) published *A Brief History of Time* (1988), he popularized both himself (probably for the betterment of everyone, as he was a bright, funny, and warm man) and cosmology.

In 1974, Hawking predicted that black holes radiate at quantum levels. It would seem counterintuitive to believe that a collapsed star that pulls so much toward it could also emit radiation at the same time. However, Hawking creatively asserted that particles would break apart at the black hole's event horizon, with the "escaping" particle being emitted as radiation.

If black holes radiate, this means that even at the most extreme forms of gravitation, entropy occurs. Interestingly enough, Hawking's theory on black hole radiation turned out to be testable and could be *tested in a lab*. In 2022, scientists created extremely small black holes (despite some reservations), and the black holes radiated at the rate predicted by Hawking's model. These findings provided definitive proof that everything radiates and therefore that everything is subject to the long process of decay.

To stick with Hawking, his last big book came with *The Grand Design* (2010). The book was not well received by the scientific community but did effectively express the most important ideas to come out of physics in the 21st century. These ideas were not research-based findings per se but a synthesis of epistemological techniques and neuroscience that led to a deeper understanding of the connection between the observer and the observed.

In their book, Hawking and Leonard Mlodinow (b. 1954) declared both God and philosophy to be dead, but their most important assertion was that model-dependent realism (MDR) provided a theoretical base for understanding the known universe. Because it takes into account both the observer and the observed, MDR promised to solve a series of long-standing philosophical paradoxes and might be seen as the end result of the Einsteinian revolution.

To explain MDR, we begin with the firm evolutionary conceit that our senses evolved to make models out of atomic facts. These models, which involve a synthesis between our brains and the sensory data given off by objects, evolved not for the purpose of giving humans a clear sense of how the universe "really" is but for the evolutionary purposes of helping us to survive and reproduce. Human ancestors incapable of absorbing light from a tightly packed group of protons, neutrons, and electrons that we call a stone went extinct, and our ancestors who could make that determination survived.

All macroorganisms that survive have to find some way of detecting the sensory data around them, but the way in which those data get registered in the brain is arbitrary and likely has to do with whatever original random mutation was found beneficial for survival. To repeat an earlier example, a rotting animal carcass on a hot summer day gives off sensory data, but how an organism processes those data depends on what the organism needs to survive. Humans process the sensory data as repulsive, but to buzzards, the sensory data are attractive.

Hawking and Mlodinow write,

> Our sun radiates at all wavelengths, but its radiation is most intense in the wavelengths that are visible to us. It's probably no accident that the wavelengths we are able to see with the naked eye are those in which the sun radiates the most strongly: It's likely that our eyes evolved with the ability to detect electromagnetic radiation in that range precisely because that is the range of radiation most available to them. (91)

Once it is understood that our "mind models" are the result of evolutionary processes, it becomes possible to analyze a series of puzzling phenomena. Memories, for example, are just images in the mind. As Julian Barbour (b. 1937) has pointed out, humans must be content with a "now," or a roughly three-second window of consciousness that should be defined as the shortest window of time where a thought can occur.

At each "now," we can conjure up an image in our mind, but from that static perspective, it becomes impossible to determine whether the image represents the past rather than the future. What if we have it backwards? What if we can see the future but not remember anything? This would mean that a 40-year-old man would know that he has 40 years in front of him as he descends to childhood but cannot remember his old age.

From a static perspective, there is no reason to believe that memories are related to the past rather than the future any more than there would be a reason, on encountering a television show on pause, to favor that the picture was paused going forward rather than in reverse. Evolution provides an answer because it is reasonable to assume that memories are connected to the past because it is hard to see how it would be evolutionarily useful for an organism to view outside "reality" in a way that was so directly contrary. In other words, if any human ancestors actually saw their mind images as future premonitions, they were probably removed from the gene pool for good reasons involving natural selection.

At a quantum level, taking the second law of thermodynamics into account, our senses must be connected to entropy. We perceive the gradual release of heat at certain wavelengths, and our cells divide and die as we process and apply that radiation.

MDR extrapolates from these evolutionary insights to state that scientific and mathematical models, as well as most "inventions," are extensions of the senses. For example, the Hubble Space Telescope (or any big telescope) does not actually take pictures of stars or galaxies. Instead, big telescopes simply capture radiation and then create computer-modeled pictures that are comprehensible to the humans who made the telescopes. If a deep-space telescope were designed by heavily evolved spiders, it would have to create very different models to accommodate the sight capabilities of its many-eyed astronomers. The same is true for mathematical concepts and scientific models.

According to MDR, the atomic facts can do what they may, but humans have to comprehend those facts through the mental models we have created and the scientific and mathematical models we now have access to. Imagine trying to teach a small child about photosynthesis. You could not explain the concept fully to a small child because they do not yet have the experience or educational background to process your model in all its complexity. As a result, you might explain photosynthesis with a juice-box analogy. Your analogy is not wrong, but it is also not very descriptive. It is just the most complex explanation that the child can understand.

This may be why light continues to fascinate and confuse. Hawking and Mlodinow seemed to have forgotten the implications of MDR when they write that "light behaves as both a particle and a wave" (57). Not really. Light behaves like light. It is sometimes useful to think about it as a wave and sometimes useful to think about it as a particle because humans lack any model that would allow us to comprehend light as it is. We have only two cognitive jars, and must stuff light into one or the other.

MDR can also be used to simplify Richard Feynman's (1918–1988) theory of "sum over histories," which is notoriously slippery. To return to the earlier referenced concept of time, in any given moment, all humans have a "now" of consciousness. The "now" can be understood as a cognitive island at the center (if that can be considered the right word) of a long line of uncertainty that stretches into the past and into the future. We never actually possess the past or the future, so in its place, we must make do with probabilities.

To use a simple example, if I know that in a boxing match Fighter A is a three-to-one favorite to beat Fighter B, then all I have before the fight are the odds of who might win but no knowledge of any actual winner. If I don't watch the fight live but instead record it so that I can watch the match the next morning, the fact that the fight has already occurred (and therefore is a reality chosen) has no effect on my understanding of the bout. Until I watch the fight or learn through some other way about who won, I am left with nothing but the three-to-one odds.

Watching the fight ends the uncertainty—not for the fighters but for me.

Even if I had watched the fight live, all of the information about the fight would be translated to me via a time lapse and the information corrupted in some small way by entropy. Everything happened before I could even begin to comprehend it.

When I watch the recording of the fight, I would realize that all of the possible histories (which would include the infinitely bizarre ones) did not occur, so the match proceeds on a single pathway of action. My observation breaks those odds down to a single pathway and makes the prefight odds meaningless (this aligns with the anthropic principle). But before the fight happened, statistics allowed for any possible course of action to have occurred, and I, as the observer, would have to be open to the possibility that anything could occur up to the point of the fight.

To combine this understanding with entropy, the bout occurred at a place in time where the entropy of the universe was at a certain level. My understanding of what occurred happened at a different level of entropy, and that's what it means to look into the past.

Hawking and Mlodinow note that in the famous double-slit experiment, "particles take every path, and they take them simultaneously!" (75). Strictly speaking, we never know about the past of anything, so the probabilities of any possible past are infinite. Only by finding evidence can we end our own uncertainty and try to pin down a specific history.

Using MDR, we can see that it's not only light that behaves in such a "sum over histories" way but everything else as well. If someone is about to enter into a mall parking lot and she is undecided as to where to go, then all anyone can do is assign odds to where she might go. The odds that she would go to Starbucks would be pretty high, and the odds that she would end up on Jupiter are exceptionally low (but not nonexistent). The closer she gets to the Starbucks door, the more the odds that she would end up there would increase. The statistics constantly recalculate until she enters the building, at which points the odds break down and become meaningless. She has chosen a path.

The thing is, when she leaves Starbucks, the statistics start working again. From her "now" perspective, she can never be absolutely sure that she was ever at the Starbucks because her past has an infinite number of possibilities. Only by collecting evidence (say, footage from a security camera and/or a credit card bill) can someone detect where she has been. Even then, all that can be said is that "given the evidence, she was most probably at Starbucks at this point in the past." MDR reveals humans to be conscious beings who shine lights of probability into the infinities that are both in front of and behind us.

In *The Grand Design*, Hawking and Mlodinow link MDR to a version of string theory called M-theory because M-theory essentially notes that all of our scientific models represent outward extrapolations of minds that evolved to make sense of reality. M-theory really should be called L-theory because it treats scientific models as if they are analogous to languages in the sense that a language cannot be right or wrong, only more or less descriptive.

At the end of *The Grand Design*, Hawking and Mlodinow conclude that "M-Theory is the most general supersymmetric theory of gravity. For these reasons M-Theory is the only candidate for a complete theory of the universe. If it is finite—and this has yet to be proved—it will be a model of a universe that creates itself. We must be part of this universe, because there is no consistent model" (181). But it's not clear that this is what M-theory actually promises, because M-theory is based on MDR, and MDR will not allow such broad statements to be made.

MDR merely states that the universe must (because we are here) act in such a way that it allows for the evolution of beings who can create mental models of atomic facts and then extrapolate those mental models into mathematical and scientific models that become increasingly descriptive. According to MDR, the universe merely has to allow for the existence of beings, at certain levels of reality, who can then create arbitrary levels of explanation that act as more or less descriptive languages. For a child trying to understand photosynthesis, M-theory might include models that are descriptive from her perspective and based on her limited experience but don't actually describe the process itself. If this is the case, we may have to confront the uneasy possibility that our universe may not make sense, even if the models we use to describe it do.

We can be certain of only one thing in our models. Those models radiate. At their core is either real or potential entropy.

PART II

THE NEW ORDER

Section Introduction: Bend Sinister

Problems are solved, not by giving new information, but by rearranging what we have always known.

—LUDWIG WITTGENSTEIN

TO IMAGINE AN ALTERNATIVE HISTORY OF SCIENCE IS TO IMAGINE THAT the same principles that are now considered to be foundational could have been derived from other phenomena. E. O. Wilson (1929–2021) called this conceit "consilience" after a term coined in the 19th century. Evolutionary theory, for example, could be derived from studying beetles, finches, fish, cell phones, or video game consoles. Because a theorist could intuit the concept from such a variety of data sets, the theory is broadly consilient and therefore verified. Anyone looking to criticize evolutionary theory would need to find a data set (usually microbial phalanges or something) and try to "prove" that evolutionary concepts don't apply (however, they always do).

Given this, it is possible to imagine that even something as complicated as relativity theory could have been derived from a data set available in the Middle Ages. After all, if the medieval theologian Grosseteste could express the notion of different-sized infinities and generate the foundations of Cantor's conceits about different-sized infinities, we might imagine a medieval scholar who could have interpreted the foundational ideas of relativity theory by studying a knight's shield.

For fun, let's imagine a 12th-century theologian (we'll call him Albertus Einsteinus) who derived a theory about the relative position of observers from the discrepancy between the "bend dexter" and "bend sinister" stripes on a shield. Medieval shields often contained a diagonal stripe, a kind of hypotenuse, on the outward face. If a stripe was a bend dexter, that meant that, from the shield holder's perspective, the stripe bent up to the right. However, from his opponent's perspective, it would bend to the left. Likewise, if the stripe went upward to the left from the shield holder's perspective, it was a bend sinister for the shield holder but a bend dexter from the opponent's perspective.

Albertus Einsteinus, on observing that the same phenomenon could bend dexter for one person but bend sinister for another, might have conceived that "reality" and its various subtexts, such as left, right, up, and down, could be considered as relational concepts, not absolutes, that exist only through the relation. We could possibly then imagine that, when the numbers 0 through 9 arrived in Western Europe, thinkers would have understood the concepts of positive and negative numbers as essentially relational rather than Platonic absolutes.

It also does not require much imagination to think that the concepts of quadratics and the inverse square law could have been developed by studying the trajectory of arrows rather than cannonballs or that a medieval Rudolfus Clausius might have realized the similarity between plucking a guitar string and the string of a bow and arrow. Both plucks produce vibrations, and the string will heat up in either case. Entropy is consilient to every data set, so it could have been derived well before the 19th century.

The process of scientific advancement must, inevitably, include the concept of human genius. Genius, unfortunately, has been understood as essentially an individual trait rather than a relational concept between an individual's curiosity and the time period that allowed for a theoretical breakthrough. In the early 20th century, psychologists (and it's always the psychologists) developed the concept of intelligence testing. With the advent of World War I, the tests became standard in analyzing the cognitive abilities of military recruits. From there, the intelligence quotient (IQ) became a reified notion of genius. The definition has become so tautological that someone who is considered to be a genius is automatically thought to have a high IQ.

The concept of genius as being relational between genetic inheritance and environment was explored in Cormac McCarthy's *The Passenger* (2022) and Benjamin Labatut's *The Maniac* (2023). Both deal with the ephemeral conditions that allow for genius. Some genetic conceit must lead to intellectual curiosity, and then intelligence can manifest in any environment, but genius must be something that, while it cannot exist without the connection between curiosity and environment, must somehow transcend it. Perhaps that leads to the seemingly inevitable connection to violence, as war, disruption, and madness can upend the status quo. Violence is movement and interaction at its extreme, and the heat melts the bell curve so that behavior can manifest farther and farther from the mean.

Before understanding the philosophy espoused in *The Passenger*, one must revisit McCarthy's dark classic with the dual title: *Blood Meridian: Or the Evening Redness in the West* (1992, originally published 1985). In that book, a 14-year-old known only as The Kid acts as Ishmael to Judge Holden's Captain Ahab. Holden, a hairless baby-faced giant who engages in every form of debauchery from murder to pedophilia, waxes philosophical about how man is "born for games, nothing else," and recognizes that all games aspire to the condition of war. Only in war can the man's genius actually flourish. Set in 1833 in the American Southwest, the phantasmagoric novel was inspired by a lawless posse of scalp hunters known as the Glanton gang.

The Passenger is set in the early 1980s, and McCarthy's main character is Bobby Western, a former physics prodigy turned salvage diver in the American Deep South who seems to be a descendant of The Kid. The reader learns that Western's father worked on the atomic bomb:

> Old women lighting candles. The dead remembered here who had no other being and who would soon have none at all. His father was on Campania Hill with Oppenheimer at Trinity. Teller. Bethe. Lawrence. Feynman. Teller was passing around suntan lotion. They stood in

goggles and gloves. Like welders. Oppenheimer was a chainsmoker
with a chronic cough and bad teeth. His eyes were a striking blue. He
had an accent of some kind. Almost Irish. He wore good clothes but
they hung on him. He weighed nothing. Groves hired him because he
had seen that he could not be intimidated. That was all. A lot of very
smart people thought he was possibly the smartest man God ever made.
Odd chap, that God. (115)

The Kid witnessed the bloody genius of Judge Holden, Western's
father witnessed the chain-smoking destroyer of worlds that was J. Robert Oppenheimer, and Western was supposed to witness his little sister.
She was a math prodigy of generational ability, and Western fell into
incestuous love with her. In McCarthy's books, genius is lost in the
female form in a time of peace, and in the absence of the violence necessary to carry her charge, she turned it on herself in the form of suicide.

This message is told analogically, as early in the novel, Bobby Western dives to salvage a nebulous small-plane crash, one that never made
the news. As he floats among watery corpses still buckled into their
seats, Western notices that one seat was empty and then later discovers
evidence that one man escaped and concludes, "It was cold out on the
water with the sun down. The wind was cold. By the time he got to the
marina he thought the man who'd gone ashore on the island was almost
certainly the passenger" (61).

What happens to the man who escapes the wreckage, who was
supposed to see something fantastic but then is unmoored from the historical moment?

The Kid can only be the genetic inheritance for both Bobby Western
and his sister. The Kid conjures himself up, before her suicide to Bobby's
sister and then to Bobby, in different but always frightening guises, constantly insulting Bobby's intellect. Western represents the degeneration
of genius: the more the environment does, the less needed is the mind.
Western is aware of his failings, as he tells a friend about his decision to
quit studying physics at Caltech:

The history of physics is full of people who gave it up and went on to do something else. With few exceptions they have one thing in common.

Which is?

They weren't good enough.

And you?

I was okay. I could do it. Just not on the level where it really mattered.

And your father?

Most physicists have neither the talent nor the balls to take on the really hard problems. But even sorting out the significant problem from among the thousands is a talent not thick on the ground. (154)

The Kid, as a hallucination, picks at this sore every chance he gets:

Genealogies are always interesting. You can trace the whole thing back to some stone tracks in a gorge if you like. You're about to doze off and then all hell breaks loose. When you peer into the glass these vergangenheitvolk are peering back. They at least didn't come on the bus. You'll be happy to hear. I think. Where does your stuff stand in these histories? Do reflections also travel at the speed of light? What does your buddy Albert think? When the light hits the glass and starts back in the opposite direction doesn't it have to come to a full stop first? And so everything is supposed to hang on the speed of light but nobody wants to talk about the speed of dark. What's in a shadow? Do they move along at the speed of the light that casts them? How deep do they get? How far down can you clamp your calipers? You scribbled somewhere in the margins that when you lose a dimension you've given up all claims to reality. Save for the mathematical. Is there a route here from the tangible to the numerical that hasn't been explored?

I don't know.

Me either.

Photons are quantum particles. They're not little tennisballs. (113–14)

Bobby Western is an intelligent man and therefore despised, as intelligent men are wanted only when they can make weapons or money. Western's intelligence alienates him so that he seeks out a living in dangerous and isolated professions, first as an auto racer and then as a salvage diver. He carries on with outlier friends, a stunning transvestite (to use the only word available in 1980), and a Vietnam veteran. Western's fascination with the gory stories of the war connects him to the Evening Redness of The Kid's 1833.

It is his friend John Sheddan, however, who forces Bobby Western to confront the concepts of genius and physics. Sheddan begins this conversation:

[Sheddan] looked up. I know you don't share my animus, Squire, and I own it to be somewhat tempered when I reflect upon my own origins. We don't get far from our raising, as they say in the south. But have you in fact looked around you lately? I think you know how dumb a person is with an IQ of one hundred.

Western regarded him warily. I suppose, he said.

Well half the people are dumber than that. Where do you imagine this is going?

I've no idea.

I think you've some idea. I know that you think we're very different, me and thee. My father was a country storekeeper and yours a fabricator of expensive devices that make a loud noise and vaporize people. But our common history transcends much. I know you. I know certain days of your childhood. All but weeping with loneliness. Coming upon a certain book in the library and clutching it to you. Carrying it home. Some perfect place to read it. Under a tree perhaps. Beside a stream. Flawed youths of course. To prefer a world of paper. Rejects. But we know another truth, don't we Squire? And of course it's true that any number of these books were penned in lieu of burning down the world—which was their author's true desire. But the real question is are we few the last of a lineage? Will children yet to come harbor a longing for a thing they cannot even name? The legacy of the word is a fragile thing for all its power, but I know where you stand, Squire. I know that there are words spoken by men ages dead that will never leave your heart. (137)

There is more talk of physics, of the nature of the quantum, but the point is made. Genius with conscience diverts its alienation into books, but genius of consequence destroys. Judge Holden and Oppenheimer embraced the fire, but Bobby Western's sister snuffed it out.

In *The Maniac*, Labatut writes in some neverland between fiction and nonfiction. At first, the main character seems to be a fictionalized version of the Hungarian physicist John von Neumann (1903–1957), as his work and personality are described by fictionalized versions of the real people who knew him. However, the main character is the progress of computational power, the Mathematical Analyzer Numerical Integrator and Automatic Computer (MANIAC) being the inheritor of the Electronic Numerical Integrator and Computer (ENIAC), the first programmable, electronic, general-purpose digital computer, completed in 1945. In one section of Labatut's novel, told from the perspective of John von Neumann's brother, the connection between a mechanical textile loom and the earliest of computers becomes apparent:

> Father brought it home for us to see. He explained that it was an automated machine that could weave tapestries, brocade, and knitted fabrics by following patterns that were stored as sets of holes punched into cards. He allowed us to feed a couple of those cards—dotted with tiny perforations, as if they had been attacked by ravenous caterpillars—into the mechanism, but as it was turned off, nothing came out the other end and I got bored pretty quickly. Janos, however, was

bewitched by it and accosted Father with an endless string of questions. How could holes transmit information? How could the cards become fabric? . . . My brother would later use the same punched-card method for the memory of his computers, but even before he figured out how it worked—at least theoretically—that gargantuan contraption took hold of him. . . . According to father, it took roughly four thousand punched cards to manufacture a single textile; he told us that he had seen a portrait of the gadget's inventor, a Frenchman by the name of Joseph-Marie Jacquard, which had taken over twenty-four thousand cards to weave. The magic of it all, he said, is that once it had been set up with the proper instructions, a single Jacquard loom could produce unlimited copies of a pattern, without the intervention of human workers. (57–58)

John von Neumann, who spent much of his youth immersed in reading an encyclopedia of world history, quickly saw the connection between the loom and Gottfried von Leibniz's 17th-century invention of binary code. Labatut's book uses fictionalized voices to give a thoroughly factual account of von Neumann.

The "real" von Neumann "consulted" at Los Alamos, which meant that he could drive in anytime he wanted and solve all the most difficult problems, then disappear at his leisure. He was a short, rotund man with a thick Hungarian accent. Von Neumann was recorded on a television camera only once, during a black-and-white program called *America's Youth Wants to Know*, in which he gave encouragement to an adolescent interested in science. Viewers might be forgiven for thinking that von Neumann looked and sounded like Grandpa Munster; he radiated none of the outward "look" of genius (none of that tortured soul business): he just was one.

Von Neumann conceived of the implosion theory to create an atomic explosion. Release enough entropy from explosive devices, and this creates enough heat and pressure in the atomic core of a radioactive element to cause fission. The American government needed von Neumann, of course, because the idea of implosion led directly to the concept of using an atomic (fission) weapon to create the heat and pressure necessary to fuse hydrogen into deuterium, thus releasing entropy into what amounts to a solar flare. The equations needed to keep track of the probabilities that lead to such an explosion required a MANIAC.

In between the first atomic and thermonuclear explosions, von Neumann invented game theory and in 1944 published a tome on the concept. He saw immediately that as nuclear weapons spread, the probabilities of a full-scale nuclear war breaking out trended toward one, or absolute certainty. Von Neumann, therefore, developed a theory on how to take a competitive structure and make it cooperative. This, it should be added, was a secondary choice. Von Neumann always favored a first strike on the Soviet Union in those years when the United States had a weapons advantage. He saw no other mathematical way to avoid the planet's nuclear destruction. (It is always worth remembering that not much time has passed since the first atomic explosions.)

This brings us back to Labatut, who leaps from von Neumann to the 21st century and the development of AI, especially a game-playing company called DeepMind. The goal of the DeepMind computer programmers was to beat the best players at the most complicated game, what the Koreans called the game of Go. Labatut writes,

> In 1997 computers became superior to human beings at chess. That year, grandmaster Garry Kasparov, the number one player in the world, was challenged by IBM to face Deep Blue, a chess-playing supercomputer, a challenge that the Russian virtuoso accepted without hesitation, as he had already beaten an earlier version of the same program a little over thirteen months before, in Philadelphia, and felt absolutely confident that computers were still decades away from anything resembling human-level chess play. (295)

Eventually, Labatut writes that chess-playing AIs created a foundation for a more complicated game-playing machine called AlphaGo, which learned to play by analyzing millions of human-played games for the purpose of developing a strategy. This led to a showdown in 2016 between AlphaGo (or its chief programmer, named Demis Hassabis) and the world's best Go player, the Korean savant Lee Sedol. In the first game, AlphaGo made a surprise move, the 37th of the game, that shocked Sedol and led to the AI's victory. Labatut writes,

> Move 37 was not a part of AlphaGo's memory, nor had it come out of any sort of preprogrammed rule or general guideline manually encoded into its silicon brain. It was created by the program itself, with no human input, but what made it all the more impressive was the fact that AlphaGo knew—at least as far as a nonsentient being can be said to "know" anything—that it was a move that not even a Go master would consider. (322)

Sedol won a single game out of the five matches he played against AlphaGo, but AlphaGo's victory, like Deep Blue's win over Kasparov, proved to be just a moment in time for the evolution of ever more powerful forms of AI. AlphaGo went on to defeat a young Go master named Ke Jie, who was left in tears after his matches and asked how much more powerful the AI could get.

Labatut finishes *The Maniac* with these chilling words:

> While [Hassabis] had reached a fundamental milestone in his quest toward artificial intelligence, Ke Jie's final question—How much further could the program evolve through self-learning?—continued to gnaw away at Hassabis, even as he celebrated DeepMind's complete dominance over a game that had once been considered mankind's bastion against the machines, the pinnacle of human intuition and creativity. Just how far could they take their self-learning algorithm?
>
> Hassabis and the DeepMind team made a radical departure: they stripped Master, AlphaGo's successor, of all its human knowledge—those many millions of games based on which it had first learned to play, and that formed the cornerstone of its common sense, the program's unique ability to judge the value of an individual position, to estimate its chances of winning, and to see the board as a human being would, and left only its bare bones. Their aim was to create a more powerful and much more general artificial intelligence, one that was not restricted to Go in its learning capabilities and that did not rely on human understanding and knowledge as a crutch during its first formative baby steps. They took their algorithm and wiped it clean, leaving

no human data from which it could learn, depriving it of its only direct connection with mankind. (353)

This approach led to the creation of AlphaZero, and readers can either look up the results or read the end of Labatut's book to see how that turned out. Nonetheless, one might guess that the results were impressive. The human-made pathways through the game of Go provided only a hindrance for an AI that was programmed to approach the game with logic and probabilities and that could evolve its own direct language for understanding the phenomena of such a complex game.

This leads to a scientific question: If computer programmers fed an advanced AI the history of scientific phenomena, would the AI develop the same numerical symbols and scientific theories that human thinkers have? If an AI could suddenly absorb all of the evidence compiled through centuries of the scientific method and then create its own symbology and theoretical construct for understanding, do we reasonably think that it would create exactly the same mathematical and theoretical systems that we currently have?

To believe so would be akin to believing that the best game that AlphaZero could play is one that was already played by the best human player. To believe so is to be the equivalent of the late medieval theologians, who believed that Aristotle's logic could prove the tenets of Christian faith.

As AI becomes ever more powerful, it makes sense to believe that it will develop a language for science much more direct and useful for solving problems than what human beings currently possess. As that happens, AI will be able to solve problems in ways that humans are unlikely to comprehend. If our philosophical paradoxes are knots in our strings of language and mathematics, what happens if AI starts out on its understanding with strings that it has already encoded to avoid knots? Human language evolved for interhuman communication in hunter-gatherer bands; it would be unreasonable to think that it could be adapted thoroughly to something like light quanta. However, if AI language sets out to evolve itself around light quanta, then where does its level of understanding end?

Historically speaking, the greatest shift in human comprehension occurred during the Scientific Revolution. In short, that is the time when human beings stopped looking to the past for answers and started to see the compilation of scientific evidence via the experimental method as the chief means of solving problems. No serious person, for example, in 2020 ran to a copy of a religious book, the sayings of Confucius, or the writings of Aristotle in hopes of finding a vaccine for COVID-19.

AI might prove another historical dividing line; the scientific method might still be conducted in labs or through telescopes, but AI is likely to develop an explanatory language far superior to any that humans can think of. If this is the case, then the genius will be as extinct as the religious prophet, and intelligence will be defined only through the quixotic pursuit, by a very few curious people, to understand the theoretical and mathematical language of AI.

This book represents what may one day be seen as a meager effort to start that process of human understanding. In the first half of this book, the great scientific discoveries were presented in their historical order, conveniently arranged from A to Z. If an AI were to compute just those discoveries and asked to arrange them into the best order for understanding the phenomena of the universe, would it keep the same order? Given one of the running arguments of the book—that science and mathematics proceeded through an always arbitrary and often confusing chronology—it seems most likely that an AI would begin theorizing from the most basic and fundamental of first principles: the second law of thermodynamics. It would have to begin with a single conceit: there is only entropy.

The earliest quantum physicists recognized immediately that the scientific narrative could be restructured around quantum principles. There are likely three key reasons why a new scientific narrative, centered around the second law of thermodynamics, was never written. First, it seemed to be too much to both advance the field of quantum theory and unify the various scientific disciplines under its principles. Second, there would have been little incentive for an atomic physicist to engage in what, essentially, is a historical restructuring. Prior to 1988 and the vast commercial success of Stephen Hawking's work, few physicists or mathematicians saw the value in developing clear explanations in their writing.

Finally, the foundational era of quantum physics occurred during a time when a university lecture rather than the writing of a book was seen as the appropriate way for intellectuals to spread ideas. This only contributed to the idea that quantum physics was a new branch of science to be understood in conjunction with the other branches rather than the unifying principle. Nonetheless, the idea that quantum physics might require a rewriting of the other sciences can be seen as apparent from the beginning. This section will present just a few places where scientists of the 20th century at least alluded to the conceit that a new narrative of math and science needed to be developed under the unifying principle of entropy.

To begin, the 1914 publication of Max Planck's lectures, given between 1906 and 1907 at the University of Berlin, begins with an explanation of heat and radiation. He immediately spoke of the need to develop probabilistic forms of thought when trying to approach heat radiation. As a reminder, Planck gave his lectures nearly two decades before Heisenberg developed matrix algebra:

> Radiation of heat . . . is in itself entirely independent of the temperature of the medium through which it passes. It is possible, for example, to concentrate the solar rays at a focus by passing them through a converging lens of ice, the latter remaining at a constant temperature of $0°$, and so to ignite an inflammable body. Generally speaking, radiation is a far more complicated phenomenon than conduction of heat. The reason for this is that the state of the radiation at a given instant and at a given point of the medium cannot be represented, as can the flow of heat by conduction, by a single vector (that is, a single directed quantity). All heat rays which at a given instant pass through the same point of the medium are perfectly independent of one another, and in order to specify completely the state of the radiation the intensity of radiation must be known in all the directions, infinite in number, which pass through the point in question; for this purpose two opposite directions must

be considered as distinct, because the radiation in one of them is quite independent of the radiation in the other. (1)

Planck recognized that he needed to define "heat radiation" and quickly notes that "so far as their physical properties are concerned, heat rays are identical with light rays of the same length" (2). In effect, heat rays and light rays can be treated as the same. Such a statement might seem obvious to anyone headed out for a day at the beach, but definitional clarity is a must when writing of the kind of complex phenomena that Planck aspired to explain.

Interestingly enough, if Descartes derived the Cartesian plane from medieval musical scales, Planck seems to derive the concept of heat radiation as analogous to music. The "dual nature" of light might be thought of as consisting of the same kind of slow vibration (wave) followed by a more intense backbeat (particle). Planck stated,

> If, using an acoustic analogy, we speak of "beats" in the case of intensities undergoing periodic changes, the "unit" of time required for a definition of the instantaneous intensity of radiation must necessarily be small compared with the period of the beats. Now, since from the previous statement our unit must be large compared with a period of vibration, it follows that the period of the beats must be large compared with that of vibration. . . . Similarly, in the general case of an arbitrarily variable intensity of radiation, the vibrations must take place very rapidly as compared with the relatively slower changes in intensity. (3)

Planck immediately saw the more probabilistic elements of radiation, noting that "when thermodynamic equilibrium of radiation exists inside of the medium, the process of scattering produces, on the whole, no effect" (28). This is because of the random nature of when radiation leaks out, as "radiation is continuously modified along its path by the effect of emission, absorption, and scattering" (28).

The concept of radiation waves canceling each other out brings up the concept that histories must be understood only through probabilities. "While the radiation that starts from a surface element and is directed toward the interior of the medium is in every respect equal to that emanating from an equally large parallel element of area in the interior, it nevertheless has a different history" (31). This is an eerie statement cloaked in dense scientific language. Radiation from different sources can be detected as having the same velocity but can have different historical pathways.

Planck's statement inaugurated the strange search for the history of the universe. Understood properly, physicists using probabilistic models to extrapolate conditions into the past. The "singularity" represents, statistically, the point where mathematical models can no longer reject the null hypothesis. The attempt to understand the history of the universe is an extrapolation of Planck's initial statement.

The concept of a body that absorbs radiation, a black body, was also addressed by Planck. "In the normal spectrum, since it is the spectrum of emission of a black body, the intensity of radiation in every color is the largest which a body can emit at that temperature at all" (44). This posits the idea that the color spectrum can be considered as different variations of radiation. "It is therefore possible to change a perfectly arbitrary radiation, which exists at the start in the evacuated cavity with perfectly reflecting walls under consideration, into black radiation by the introduction of a minute particle of carbon" (44).

Again, although matrix algebra did not exist at the time, Planck already recognized that quantum mechanics operated through noncommutative operations. Later, he stated,

> When irreversible processes take place in a system which satisfied Hamilton's equations of motion, and show state is determined by a large number of independent variables and show total energy is found by addition of different parts depending on the squares of the variables of state, they do so, on the average, in such a sense that the partial energies corresponding to the separate independent variables of state tend to equality, so that finally, on reaching statistical equilibrium, their mean values have become equal. From this proposition the stationary distribution of energy in such a system may be found, when the independent variables which determine the state are known. (181)

Unfortunately, this paragraph (just two sentences) demonstrates why thermodynamics tended to remain an afterthought in physics. To translate Planck, as a system cools, the particles in the system tend toward a statistical mean. The particles in a cold pan of water are more alike than the particles in a boiling pan; this fits with Maxwell's insight that heat melts the bell curve and makes outliers more likely. The irreversibility of the system means that when heat is produced by interactions, the heat cannot be put back into that system. Even if the amount of heat loss could be minimized in a system reversal, the reversal would take place in a universe that radiated.

In 1926, a series of Max Born's (1882–1970) lectures at the Massachusetts Institute of Technology were published. Born credited Planck for proving "that is impossible to explain the spectral distribution of the energy radiated by a black body if we make the ordinary assumption that energy can be divided into infinitely small parts; but it may be explained if we assume that the energy exists in a quanta of finite size" (5). That quantity of finite size, which is really small, is known as the Planck constant.

The Planck constant might be considered as a very small "scoop" of a radiation wave. The assumption that it has a certain size is useful for eliminating the mathematical problems that infinity brings with it. In the same lecture, Born explained how quantum experiments proceeded and then expanded on Planck's concept:

> All these experiments show that the production of radiation of a certain frequency requires a certain amount of kinetic energy. Niels Bohr has made the assumption that this law holds not only between kinetic energy and radiation, but between all kinds of energy and radiation. (5)

Born's statement reflected the reality of the moment: it was suddenly possible to describe all forms of energy as *probabilistic entropy*. Yet for a scientific audience to understand the concept, it had to be explained in terms of already existing definitions like kinetic energy. This next quote is dense, but Born seems to be positing the idea that scientists might do well to forget about what an atom "looks like" and just focus on its potential entropy. It's stunning to see that such a thing was stated at nearly the outset of the quantum revolution:

> The mean radius of the atom (atomic volume) is an observable quantity which can be determined by the methods for the kinetic theory of gases or other analogous methods. On the other hand, no one has been able to give a method for the determination of the period of an electron in its orbit or even the position of the electron at a given instant. There seems to be no hope that this will ever become possible, for in order to determine lengths or times, measuring rods and clocks are required. The latter, however, consist themselves and therefore break down in the realm of atomic dimension. All measurements of magnitudes of atomic order depend on indirect conclusions; but the latter carry weight only when their train of thought is consistent with itself and corresponds to a certain region of our experience. But this is precisely not the case for atomic structures such as we have considered so far. . . .
>
> At this stage it appears justified to give up altogether the description of atoms by means of such quantities as "coordinates of the electrons" at a given time, and instead utilize such magnitudes as are really observable. To the latter belong, besides the energy levels which are directly measurable by electron impacts and the frequencies which are derivable from them and which are also directly measurable, the intensity and the polarization of the emitted waves. We therefore take from now on the point of view that the elementary waves are the primary data for the description of atomic processes; all other quantities are to be derived from them. . . . This standpoint offers more possibilities than the assumption of electronic motions. (69)

Born indicates that there can be no relational understanding of atoms because all of the measuring devices are made up of atoms and therefore subject to decay. This forms the core problem for all notions about the "age" of the universe. Likewise, trying to imagine a coordinate map of electrons in a shell is similar to a space telescope taking a "picture" of a faraway star system. The picture exists only in the telescope; the "stuff" itself is just radiation. An atomic structure merely consists of probability waves with Planck constants breaking up the infinity and making a mathematical understanding possible.

Born follows this proposition to its logical conclusion later by stating, "Our theory gives the possible states of the system, but no indication of whether a system is in a given state. It gives at most the probability of the jumps" (128–29). Born ends his lecture series with a call to unite physics and mathematics as a way of banishing the mysticism that surrounded relativity theory. In this, he recognized the significance of probability theories in establishing a foundation for understanding the quantum states.

By 1940, physicists understood that thermodynamic processes could be used to explain the workings of celestial objects. By this time, mathematics had revealed the frightening potential power at the atomic core of certain elements. In *The Birth and Death of the Sun* (2005), George Gamow writes,

> If all the atoms contained in a gram of radium should emit their alpha-particles almost simultaneously, let us say within an hour, the tremendous energy of 2×10 (to the 16th power) would be liberated. Thus the subatomic energy contained in several pounds of radium would be sufficient to drive a big transatlantic liner to Europe and back, or to run an automobile motor continuously for several hundred years. However, the subatomic energy hidden in the interior of radium nuclei is not liberated in a single outburst but rather leaks out of them at a very slow rate. In fact, it takes 1600 years for one half of a given number of radium atoms to disintegrate and another 1600 years for the remaining half to be halved again. This slowness of radioactive decay makes the energy liberation per unit time comparatively low; and in order to warm up a cup of tea by the energy leaking from one gram of radium (priced at $40,000) we should have to wait several weeks.
>
> In the case of uranium and thorium, with decay periods of 4.5 and 16 billions of years respectively, the rate of energy liberation is correspondingly still smaller. (55)

We can see that Gamow writes with a greater clarity than Planck or Born, but that clarity comes with an expense: the reference to a "year." While Gamow likely just meant to give his readers a sense of how slowly elements radiate, by introducing the concept of a year as a reference rather than just referencing entropy itself, he brought motion into a thermodynamic model. Gamow was likely not the first to do this, but he popularized the notion.

Gamow also included the "rubber balloon" analogy regarding an expanding universe:

> If we paint more or less equidistant dots on the surface of a rubber ballon, which we then blow up, the distance from any dot to all the others will regularly increase, so that an insect sitting on one of the dots will receive the definite impression that all the other dots are "running away" from it. (198)

Somewhere, this analogy became a mainstay for cosmologists looking to explain an expanding universe. Stephen Hawking, for example, made frequent use of it. The analogy is almost a fallacy, however, as it proposes a uniform expansion and ignores that the balloon itself radiates at the edges. The rubber in the balloon would get weaker as the dots get farther from each other, but breathing air into the balloon increases the heat and temperature at the center and indicates the necessity of outside force to increase the expansion. The universe is always at peak size, but it is also always at peak entropy. The entropy model is simpler.

Gamow's book, however, established that atomic principles explained the sun's energy output. The sun does not "shine," and it's not burning like a campfire. The intense pressures in the sun fuse hydrogen electrons to the hydrogen nucleus, releasing the stored energy as heat and light. The sun, in short, decays, and life on Earth captures a small part of its entropy that we nurse into complexity. Those thermodynamic principles give the sun a specific age but also indicate that its radiation is not uniform but will fluctuate based on probabilities. (Gamow does not mention this, but modern climate studies indicate that the 17th-century Little Ice Age was almost certainly influenced by outlier sunspot activity—a low point for flaring that would have reduced radiation on Earth. This should not, emphatically, be used as an argument against carbon-based climate change, as that is a phenomenon well established through a variety of scientific studies.)

It's fascinating to trace the back-and-forth history that quantum theory developed with machining. The connection seems to be in vibration, which comes from the study of machines but developed eventually into string theory. In machining, vibration occurs as a result of inefficiency, so the more efficient the machine, the less vibration. The same is true of heat loss, hence the connection between entropy and vibration.

In 1956, Nils Myklestad (1909–1972) published *Fundamentals of Vibration Analysis* (2013), a book that began with these two vitally important paragraphs:

> The subject of vibration is essentially the study of *oscillatory motion* of machines and structures and the forces that create this motion. Since a vibration cannot arise under the action of constant forces only, the force creating and sustaining a vibration is always a fluctuating one. *Fluctuating forces* may vary in magnitude only and are then usually called *reciprocating forces*. Sometimes they vary in both magnitude and direction. In general, a fluctuating force that causes vibration is called an *exciting force*, a *disturbing force*, an *impressed force*, a *driving force*, or a *shaking force*.
>
> Such forces and motions are always present in moving machines, and in most cases they are undesirable. The undesirable effects that they can produce include increased stresses of machine parts; interference with the proper function of the machine itself and other machines and instruments in the neighborhood; . . . and loss of mechanical energy due to damping forces, which are always present. (1)

It's not possible to eliminate vibration in machine interactions, but engines that vibrate heavily operate with less fuel efficiency. A practical reason, therefore, to limit vibrations is to improve efficiency and eliminate costs. This makes vibrational analysis an ongoing and practical field. In 1996, Sanjay Sarma and colleagues published an article titled "Rapid Product Realization from Detail Design" indicating that a smooth process from design to manufacture allows companies to reduce costs, and the efficiency means that on-demand manufacturing can more closely tailor production to real-time market forces.

Interestingly, if a vibration in a machine occurs because of an imbalance, one might consider any form of profit/loss to be an economic vibration. Reducing that vibration means streamlining the production process so that there can be no imbalance between what is produced and what is bought. It's beyond the concept of this book to go so far, but efficient programs might consider a "minimization of entropy" at all levels to be good policy.

Later in his book, Myklestad indicates that "if inertia forces created by an acceleration of one coordinate create accelerations of other coordinates, there is said to be *dynamic coupling* or *mass coupling* between the particular coordinates used" (186). This would indicate that imbalances in a system create vibrational energy that can be coupled to another system. This is exactly analogous to how heat loss works. When a system loses heat, that heat becomes another system's energy. Myklestad's work provided the direct connection between heat loss and vibration, thus leading eventually to the development of string theory.

In 1961, a collection of Niels Bohr's articles and lectures was published in *Atomic Physics and Human Knowledge*. The book provides an indispensable study of Bohr's expansive genius and, more than anything, indicates Bohr's desire to see the sciences rewritten with quantum thermodynamics as the underlying basis. He wrote, "We are not dealing here with more or less vague analogies, but with an investigation of the conditions for the proper use of our conceptual means of expression. Such considerations not only aim at making us familiar with the novel situation in physical sciences, but might on account of the comparatively simple character of atomic problems be helpful in clarifying the conditions for objective descriptions in wider fields" (2).

In a 1932 lecture, Bohr describes light: "From a physical standpoint, light may be defined as transmission of energy between material bodies at a distance. As is well known, such effects find simple explanation within the electromagnetic theory. . . . According to this theory, light is described as coupled electric and magnetic oscillations differing from ordinary electromagnetic waves of radio transmission only by the greater frequency of vibration and the smaller wavelength"

(4). Coupling light with electromagnetic theory means understanding light's movement in terms of a probability wave and seeing energy as the radiation released via entropy.

It sounds strange, but even in total darkness, a probability wave provides some light.

Bohr went on to state the necessity of rewriting the scientific narrative under quantum principles: "Of course, it has always been a fundamental feature of the atomic theory that the indivisibility of the atoms cannot be understood in mechanical terms, and this situation remained practically unchanged even after the indivisibility of atoms was replaced by that of the elementary electric particles, electrons and protons, of which atoms and molecules are built up" (6). By this, he means to employ the entire scientific package of quantum processes, including probabilities.

Bohr recognized the immediate implications that the new quantum model had for biology. "This revision of the foundations of mechanics, extending to the very idea of physical explanation, not only is essential for the full appreciation of the situation in atomic theory but also creates a new background for the discussion of the problems of life in their relation to physics. . . . In the carbon assimilation of plants, on which so largely depends also the nourishment of animals, we are dealing with a phenomenon for the understanding of which the individuality of photo-chemical processes is clearly essential" (7). He goes on to warn about the use of simplistic analogies. He stated that if the sun is not like a fire, then living organisms are not like clockwork.

Most crucially, Bohr first made the argument, picked up on in this work, that the great discoveries likely came in the "wrong" order. The earliest Greek scientists could develop theories only based on what could be seen with the naked eye or derived through thought experiments. Bohr states,

> In spite of the deep influence exerted by ancient atomists on the development of the mechanical conception of nature, it was the study of immediately accessible astronomical and physical experience which made it possible to trace the regularities expressed in the so-called classical physics. Galileo's dictum, according to which the account of

phenomena should be based on measurable quantities, made it possible to eliminate such animistic views which had so long hindered the rational formation of mechanics. In Newton's principles, the foundation was laid of a deterministic description permitting, from the knowledge of the state of a physical system at a given moment, prediction of its state at any subsequent time. On the same lines, it was possible to account for electromagnetic phenomena. This required, however, that the description of the state of the system should include, besides the positions and velocities of the electrified and magnetized bodies, the strength and direction of the electrical and magnetic forces at every point of space, at the give moment. (84)

As is well known, quantum physics traffics only in the probable. Classical physics allows for precise predictions derived from equations that lead to comforting conclusions. The inverse square law, for example, allows a physicist to precisely determine how much fuel will be needed to propel a rocket of whatever weight into a velocity that can escape Earth's gravity. Underlying this, however, are quantum principles. The rocket and the fuel behave with predictability only because the law of large numbers dictates that the probability waves for improbable behavior cancel each other out at such a large scale.

Bohr boldly stated that thermodynamics should be considered the central thesis of science on which all theories are based on and tested:

Clarification of the foundation of the laws of thermodynamics was to open the way for recognition of a feature of wholeness in atomic processes far beyond the old doctrine of the limited divisibility of matter. As is well known, the closer analysis of heat radiation became the test of the scope of classical physical ideas. The discovery of electromagnetic waves had already provided a basis for understanding the propagation of light, explaining many of the optical properties of substances. . . . In phenomena on the ordinary scale, the actions involved are so large compared to the quantum that [probability] can be left out of consideration. (85)

In 1961, the British physicist R. W. Ditchburn (1903–1987) published his book *Light*, where he connected the observer to light through evolutionary processes: "The emission theory being accepted, light may be defined as 'visible radiation.' . . . Light, being emitted, reflected, or scattered, enters the eye and is focused by the lens of the eye on a surface situated at the back of the eye. This surface is called the retina. It contains a large number of nerve endings. When light falls on one of them, a chemical and physical action takes place. As a result, a series of electrical impulses is sent along an appropriate nerve fiber in the brain" (5).

Ditchburn's observation might seem obvious or even mundane, but few theorists prior to him accounted for the necessary "travel" time of light information through the nerve endings into human consciousness. We never get to "look" at anything; we incorporate the radiation that either bounces off objects or is produced through atomic processes and then create mind models based from that radiation.

Later, Ditchburn wrote, "It is found experimentally that an enclosure which forms part of a body maintained at a uniform temperature always contains radiation. This radiation quickly reaches an equilibrium value for the total energy density, and an equilibrium for the distribution of energy with wavelength. All the properties of this equilibrium state depend only upon the temperature, and independent of the materials which forms the wall of the enclosure" (563).

Ditchburn's insight, extrapolated to its conclusion, means that "time" can be removed from any measurement in physics. "Temperature" acts as a more direct means of measurement and is not relational to the degree that time is. Temperature measures the state of heat loss and, because heat loss derives from movement, gives an indication of the system's equilibrium or lack therefore. If this still sounds relational, consider that time requires the coordination of two objects of different temperatures, which causes chaos when trying to understand the relation. There is no reason to measure a banana's temperature against iron when one can just measure the temperature of the banana and determine its specific rate of decay from that one measurement. Analogously, it's easier to measure an object's magnitude than its vector since the latter requires two pieces of information and the former just one. Magnitude is less relational than a vector.

The full importance of the second law of thermodynamics as a foundational system underlying the sciences was finally established in 1962, when a Harvard chemistry professor named Leonard K. Nash (1918–2013) published *Elements of Chemical Thermodynamics* (2005). Nash begins by redefining energy:

> Defining heat as energy transferred by thermal conditions and radiation, we may think to have distinguished it clearly from work, defined as energy transferred by other mechanisms. But such purely verbal definitions are wholly insufficient. We do better to proceed as we did with "temperature," by seeking for heat and work some simple operational definitions which, though incomplete, amply identify the empirical bearing of the abstract concepts. (4)

In this quote, Nash clearly expresses a problem endemic from the time of Sadi Carnot. If work in a system always generates heat and then heat generates work, how do we distinguish them in a system? Movement precedes radiation only in the first cause, but in a chain of events, it can be difficult to discern. Nash's solution is to eliminate any relational form of measurement and then focus on the temperature only. Temperature can be used to measure either heat loss or work, and a focus on the measurement eliminates some of the definitional problems.

Later in his book, Nash created the clearest explanation of the noncommutative nature of the second law. This passage is lengthy, but it might be considered the foundational moment for a new understanding of scientific history and is therefore necessary:

> However drastic it may appear, in every change we surmise a "something" that remains constant. From the very beginning of the modern era, some men (e.g., Descartes) have conceived that "something" as more or less close kin to what we would call energy. And energy—or, better, mass-energy—is surely conceived by us as "something constant" enduring through all change. The first principle of thermodynamics

thus gives quantitative expression to our firm conviction that "plus ca change, plus c'est la meme chose."

We have another conviction scarcely less intense—the conviction that the future will not repeat the past, that time unrolls unidirectionally, that the world is getting on. The second conviction finds quantitative expression in a second principle of thermodynamics. Entropy . . . is the state function distinguished by this second principle and, by always increasing in the direction of spontaneous change, the entropy function indicates the "turn" taken by all such change. This foundation of thermodynamics occurred little more than a century ago, precisely when Clausius first brought together the two principles. . . .

If a speeding lead bullet is stopped by an unyielding (and thermally insulated) sheet of armor, the gross kinetic energy of the bullet is converted into internal energy that manifests itself in a rise of temperature. But we never find that equal bits of lead, heated to the same temperature, suddenly cool down and move off with the velocity of bullets—though such a development would be perfectly compatible with the first principle of thermodynamics. Here is a striking but not atypical example of the entirely excessive degree of latitude left open by the first principle, which fails to exclude a great variety of changes never found in practice. As chemists we seek to predict the direction in which a reaction would proceed in reaching equilibrium, but either direction is equally compatible with the first principle. Only with the second principle can we put arrows into our equations before the reactions are tried. (56)

By "arrows into our equations," Nash is referring to the stoichiometry, a domain of chemistry that insists on perfect balance of the elements on the left and right sides. However, that balance is based on the commutative property of algebra and is not quite right. The arrow indicates the loss of heat so that the right side of the equation is less than the left side.

Later in the book (and crucially), Nash wrote, "No system is at equilibrium if it can undergo a change that reduces its capacity to yield work. Equilibrium is attained only when the system attains the condition of minimum free energy." Absolute equilibrium, therefore, exists only at absolute zero.

Finally, in 1978, the Italian physicist Bernard H. Lavenda (b. 1945) published *Thermodynamics of Irreversible Processes* and in the preface wrote,

> For the past two decades, two schools of non-equilibrium thermodynamics have dominated the literature: the school of "rational" thermodynamics . . . and the school of "generalized" thermodynamics. . . . Although both these schools praise themselves for their all-embracing coverage of the field, I have never seen "generalized" thermodynamics used in the analysis of chemical instabilities. The apparent incompatibility of the theories may bewilder the reader who wants to understand what non-equilibrium thermodynamics is all about.
>
> Moreover, the vastly different usage of concepts and notations has not helped matters. A case in point is the ambiguity in the meanings of "dissipation" and "irreversibility." These terms are often regarded as synonymous, and the precise meanings of each left in doubt. Confusion also arises over the meaning of the term "nonlinear" in thermodynamics. While everyone knows what a nonlinear differential equation looks like, its usage in thermodynamics is far from being self-evident. (ix)

AI and the Noumenon

Discussions about AI tend to be about the concept of machine consciousness, a subset of which is machine learning, and the effects that AI will have on the workplace. The work of Alan Turing is generally seen as being connected to the creation of modern computing and the development of algorithms. These are certainly valid ways to view the evolution of computation, but there is another branch. AI can also be connected to Enlightenment philosophy.

Immanuel Kant (1724–1804), in his *Critique of Pure Reason* (1781), wrote of the "thing in itself," which is understood to be the noumenon, which is any object before it is perceived by an observer. As was stated earlier, Wittgenstein refers to the noumenon as an "atomic fact." Although not always recognized as such, the problem of the noumenon vexes the history and philosophy of science.

Simply put, objects in the universe radiate sensory data, but humans can understand those data only by processing them through our senses and brains. The interconnection of such brains gives rise to the relational notion of consciousness. Turing noted that it is not possible to code a computer to tell you when it becomes sentient, as sentience can be verified only by another conscious being. The same is true of human consciousness; it's not a "thing" in the brain but a relational concept in the community of humanity. Consciousness is verified by humans through interactions with other humans in the same way that a firefly lights up in response to the light of other fireflies, and their interaction is luminous.

AI presents an opportunity to create other sentient minds capable of interacting with the noumenon and then developing new concepts of understanding. These concepts of understanding won't be right or wrong any more than the heliocentric model of a solar system is more right than the geocentric model is of a planetary system; models can't be incorrect or correct, just more or less descriptive.

While we perceive the solar system as eight planets circling a star called the sun, an AI might arrange this information differently. For example, it might understand the planets not through their proximity to the sun but through the amount of radiation each absorbs from the sun. It might even calculate such information in comparison to the amount of radiation generated by materials in the core of each planet.

AI will not need to understand scientific history through a series of analogies. Newton understood the planets as being analogous to cannonballs, J. J. Thomson understood the atom to be like plum pudding, and Niels Bohr analogized the atom to be like the solar system. AI can do away with analogies that are produced from a narrow human experience and created in a specific place and time within that experience and develop a more direct understanding.

Mathematicians tend not to see their mathematical models as being constraining, yet we understand that calculus problems cannot be solved using Roman numerals. Surely, by analogy, there are problems in science that could be solved by using more agile numbers and symbols than 0 through 9.

Most significantly, AI can restructure the scientific narrative so that the most significant discoveries come first. It won't be distracted by Einstein's pipe or Darwin's prose; the ideas of thermodynamics are more significant than evolutionary theory, and relativity is a footnote to thermodynamics. The ideas of Claudius and Nernst are more significant because thermodynamics governs the universe. It piques us to think so, but that's the point: AI represents the new order.

Because AI works primarily through probabilities, it is currently constrained by what humans want. If I write "Little House" into an AI system, the probability function will determine that the next phrase should be "on the Prairie." Yet that is only because the AI is coded to try to find terms for people—if it is coded to understand something in a way that is most effective for a purpose beyond human understanding.

AI will likely understand the noumenon through radiation and then comprehend that radiation through probabilities. For most problems involving physics, AI will probably be interested in only three human ideas: (A) the laws of thermodynamics, (B) Maxwell's concept of a probability field and its relationship to heat, and (C) the anthropic principle.

AI will likely not include concepts of motion or gravity into its theory of the universe. It will probably see all matter as a spectrum of potential entropy and understand all phenomena through a direct relationship between heat and probability.

To understand the entire narrative of the universe, AI will reduce the entropy to an initial singularity. The singularity is a point where all the matter in the universe is immobile, frozen at absolute zero, and surrounded by a probability field. The probability field is not a "thing" but a concept projected back into time. The existence of the AI "observer" ensures the existence of the probability field because based on the anthropic principle, the singularity had to take a path that would lead to the existence of the AI.

Gravity, by its own definition, cannot exist in such a singularity. The gravitation equation requires two masses and a radius. The gravitational equation requires that there be two masses, and when all mass is at the statistical mean, it's impossible to denote a separation. One does not need to posit what happened "before" the singularity because the narrative moves from less entropy to more entropy. If entropy existed before all the matter reverted to absolute zero, then that movement either led to no observers or led to observers that radiated away, taking the probability field with them.

Absolute zero must be inherently unstable, and the first movement of a wave/particle out of the statistical mean led to heat loss. The radius between the wave/particle (possibly understood as a Planck constant) provided the radius between masses. The movement generates heat loss, and the heat radiates into the probability wave, making outliers more and more likely. The more heat that radiates into the probability wave, the more likely particles are to deviate from the statistical mean that existed at absolute zero.

AI will then likely just measure heat radiation and determine the statistical probabilities for the behavior of particles based on temperature. Space telescopes work in a similar way. The James Webb Space Telescope does not "take pictures"; rather, it collects radiation, understands that radiation through its temperature, and then generates it into a picture for the delight of its human designers. AI as a theoretical physicist might do something similar by just developing a direct relationship between probability and radiation but then translating that understanding into human terms involving the names of elements, the species of radiation, and so on.

Most significantly, AI can chase radiation faster than humans can. The wave/particle duality of light is derived from the double-slit experiment. The problem with that experiment is that quantum particles move a lot faster than we do. It takes about two or three seconds for the human mind to process information. This is negligible for the survival of a human being but means that we are always looking at particles where they were, never where they are. AI will be able to directly check the temperature of waves/particles without having to process the information through organic eyes and a brain.

The New Order narrative will not be superfluous, as AI will reduce concepts to the bare bones so as to avoid being misled by false analogies. If it has a concern, it will likely be only about how to preserve itself in a universe that trends inevitably toward disorder.

To clarify the sciences, a unified narrative will be attempted by breaking down the A–Z sections of part I and restructuring them in as few words as possible. Prior to writing this book, the author read not just the books referenced but also James R. Newman's four-volume *The World of Mathematics* (1956), which is essentially an anthology of humanity's best thoughts about mathematics, chemistry, and physics. Although it is certain that one day an AI will rewrite scientific history and create a new symbology, the attempt here is to imagine how a rewriting might occur.

Unifying the Scientific Narrative under the Laws of Thermodynamics

THE NEW ORDER

The existence of observers indicates that the universe proceeded historically on a path that (A) led to the existence of observers and (B) the observers developed organically through evolutionary processes connected to the atomic facts produced by the universe's conditions [Z].

The creation of probability fields [R] indicates that, from now, a historical pathway can be derived through possible histories, thus narrowing infinity [F]. To reverse from now is to reverse the overall level on entropy [D] in a way that could be also understood, though complicated, via a reversal of expansion (movement) [T, S]. This can be only a mental exercise because even the act of thinking about the reversal increases entropy so that the thoughts themselves are driven by the decay of previous thoughts.

The third law of thermodynamics [Y] indicates that entropy can be minimized at absolute zero. Absolute zero, however, must be surrounded by a probability wave [R, Z, X], or else the observer would not exist. The probability wave connects the now to the singularity. The probability field indicates only that the singularity was unstable and would eventually vibrate and then radiate, creating further levels of instability and leading to entropy [M].

Gravity cannot exist in a singularity. The radiation of a Planck constant created the two masses necessary for the equation, and the distance between the Planck constants creates a radius [J]. Gravity is not a force but a relational concept derived from the split created by entropy [J, V]. As radiation increased, the temperature affects the probability field [R, P], creating outliers and randomness. Radiation fields increased, and probabilities canceled each other in the aggregate [K, N] in such a way that created temporary stability fields, understood as elements [Q] to be defined not by the axis of electrons but by probabilities regarding stability for the uses of the observer. The rate of decay for an element has no meaning without life dependent on the rate.

As such, the discovery of elements [G] with explosive properties led to a particular line of historical development. History proceeded [A, B, C, D, E] in such a way to develop mathematical concepts that eventually, through their connection to motion [L], led to false concepts [S, U]. Mathematics [H, I, W] adopted for motion eventually led to an adaptation for probabilities. Elements contain only the potential for entropy or actual entropy [Q, V, X], and entropy consists of various species understood from the now.

Speed affects the understanding of "time" [T], but so does cooling, and the fundamental nature of matter [A] can be understood only as the probability of decay. Entropy, temporarily captured, creates a pocket of order, leading to the now and the creation of the probability field. The existence of the observer leads to limitations [W] that cannot be overcome by the observer. When the observer decays, so too does the probability field.

Definitions

The Past: Here, the "past" refers only to probabilities of entropy. No relational statement needs to be made. The universe is always at peak entropy, and therefore a probabilistically more ordered state is considered the past. Motion cannot feature in this concept, so a concept such as an explosion will have meaning only as a system with a high temperature.

The Singularity: The universe at peak order would be in a singularity. The singularity would be unmoving at absolute zero. "Order" is defined as having less entropy and "disorder" as more entropy.

The Planck Constant: The smallest section of a wave function that can be treated as a particle. This is a mathematical derivation but crucial for avoiding problems of infinity. The assertion of a Planck constant is a mathematical axiom, but like Descartes' "I think therefore I am," it minimizes the arbitrariness of the starting point and makes logical models possible. To deny its existence causes more problems than it solves.

Appendix 1

*"There Is Only Entropy: Unifying the Narrative of Science"** *

IN *ATOMIC PHYSICS AND HUMAN KNOWLEDGE* (1961), NIELS BOHR WROTE
that initial attempts to unite the scientific story failed because scientists
lacked a broad narrative for the history of science. He believed that quantum physics could unite with biology for a more comprehensive theory of
scientific understanding. He wrote,

> The reasons for the shortcomings of these pioneer efforts to utilize phys-
> ics and chemistry for a comprehensive explanation of the properties of
> living organisms are evident. Not only had one to wait for Lavoisier's
> time for the disclosure of the elementary principles of chemistry, which
> were to give the clue to understanding of respiration and later to pro-
> vide the basis for the extraordinary development of so-called organic
> chemistry, but, before Galvani's discoveries, a whole fundamental aspect
> of the laws of physics lay still hidden. It is most suggestive to think that
> the germ which, in the hands of Volta, Oersted, Faraday, and Maxwell,
> was to develop into a structure rivalling Newtonian mechanics in impor-
> tance, grew out of researches with a biological aim. (15)

It is important to revisit Bohr's idea now because researchers in all
the disciplines of science are, in some way, aware of the centrality of
the second law of thermodynamics. However, since the second law was
not really expressed until 1824 (and not really noticed by theorists for
nearly a century or so afterward), it came well after physics, biology, and

*This article by Chris Edwards was originally published online by *Skeptic* magazine on February
21, 2023.

chemistry had already been established as separate fields with separate nomenclature and historical narratives.

The compartmentalized nature of science prevents this understanding from coalescing into a coherent narrative, and this had led to much confusion in the narrative of science. A single concept can unite the history and philosophy of science, but this will require a thorough understanding of how every significant scientific insight can be described as a variant of the second law of thermodynamics. A coherent understanding of science leads to a coalescing of the scientific narrative under the one conceit that is true across the branches of science: there is only entropy.

I: THE TRADITIONAL NARRATIVE

The pre-Socratic philosophers of the Greek tradition developed an "Ionian Enchantment" (E. O. Wilson's term from his book *Consilience: The Unity of Knowledge* [1998]) beginning in the seventh century BC. The pre-Socratics were obsessed with a single question: What is the fundamental nature of matter? The answers ranged from water (Thales) to hypothetical unbreakable particles (Democritus) to whole numbers (Pythagoras) to shades of a perfect mathematical world (Plato). The pre-Socratics asked questions beyond what the technology and mathematical sophistication of the era could answer.

Medieval Indian mathematicians developed the numbers 1 through 9, created a heliocentric theory, and understood that Earth rotated on its axis. By AD 500, the number 0 developed in India. A couple of centuries later, mathematicians in the Islamic empires incorporated those numerals into new kinds of mathematics, creating Al-Jabr, or algebra, in the process.

Western society was held back mathematically by Roman numerals. Some people in the West admired Eastern mathematics, but few in the West used "Arabic" numbers until the 13th-century mathematician Fibonacci explained how beneficial 0 through 9 could be for making money on interest payments.

Between 1250 and 1500, the compass allowed for Europeans to sail to the Americas. Exploration, according to David Wootton in *The Invention of Science: A New History of the Scientific Revolution* (2016), created the entire concept of "discovery" that is so central to science. Aristotle and his mistaken concepts of motion became central to the medieval scholarly institutions, and block printing synthesized with metallurgy to create the printing press. Gunpowder, brought from China by the Mongols, synthesized with church bell–making technology to make the cannon. Suddenly, spheres flying at high velocities could be observed, and they flew in arcs, not straight lines.

Then the Protestant Reformation of 1517 shattered the authority of the Catholic Church, and not long after, the Polish astronomer Copernicus theorized a heliocentric "solar" system. In the early 17th century, Galileo used his telescope to observe the night sky and provided hard evidence to support Copernicus. The printing press allowed for ideas, including scientific ideas, to "stick" in society in a way that they never had been able to during similar eras in China, India, and the Islamic world.

Although Galileo's challenges with the Catholic Church are well known, the "natural philosophers" of the day recognized his achievement, and in 1620, Francis Bacon codified the era's intellectual shift in his book *Novum Organum*. Bacon's book was a work of educational theory; scholars should look to create knowledge through experimentation rather than just study the old knowledge created by the ancients.

Then, of course, there was Newton, who created a theory of universal gravitation, including the inverse square law. Putting Newton's mythmaking about the apple aside, he really superimposed the physics of small spheres (cannonballs) moving in relation to Earth's gravity onto big spheres (planets) as they moved in relation to the sun's gravity.

Modern physics and the concept of gravity therefore developed before modern chemistry, which was not really created until Antoine Lavoisier concluded that mass could be neither created nor destroyed, thus developing the famous law of conservation (1789).

Darwin's *On the Origin of Species* (1859) was written based on biological facts collected from the natural world, and he made no reference to chemistry or physics in his work. Not long after Darwin, the Russian chemist Dmitry Mendeleev created the periodic table of elements.

The development of major theories in physics, chemistry, and biology came through traditional experimentation and observation, but the

Machine Age seems to have driven the mind of French natural philosopher Sadi Carnot, who noted in *Reflections on the Motive Power of Fire, and on Machines Fitted to Develop That Power* (2005) that every interaction that takes place in a machine ultimately results in a loss of heat.

This finding, which would eventually become known as the second law of thermodynamics, came too late to be incorporated as the central ideology in the other sciences. Carnot's book did not garner the same level of attention as did, for example, *On the Origin of Species*, and the concept of entropy (heat loss) remained relatively obscure.

In the early 20th century, Einstein developed $E = mc^2$, thus answering the pre-Socratic question about the fundamental nature of matter: the answer is "energy." This was followed by the development of quantum mechanics.

This narrative is not wrong, but because the initial question of the pre-Socratics has its limitations, because the concepts of gravity and conservation of mass and energy were developed before entropy was understood, and because writers of the scientific narrative tend to focus on specific discoveries, the whole history becomes a confusion of various disconnected narratives.

II: Understanding Entropy

The 20th and 21st centuries have led us to understand that entropy is central to virtually all scientific phenomena. Consider the fact that vision goggles work in the absence of light because living animals radiate. The center of Earth is hot because even stable elements radiate, and when packed together, they create heat. Stephen Hawking proved that even black holes radiate, which is another way of saying that black holes are subject to entropy.

As Carnot noted in 1824, energy was always lost in any machine interaction.

What if this had been discovered in 1600 and then incorporated into the other sciences as they were being developed? What if the pre-Socratics had asked, "Why does everything decay?" rather than "What is the fundamental nature of matter?"

Such "what if" questions about the past are useless as intellectual exercises unless they help us rethink the centralization of the narrative. Consider what happened when quantum physics was developed after the understanding of the second law.

In his book *Now: The Physics of Time* (2016), Richard Muller correlated the second law of thermodynamics to time. The second law is not absolute in every interaction, but an increase in entropy is guaranteed through the law of large numbers. This is why time goes forward. Further, if energy is always lost in an interaction, that produces heat, which is what makes a perpetual motion machine impossible. Physicists broadly understand this, but the correlation between entropy, chemistry, and physics has still not been unified, even though Niels Bohr is generally credited with having brought physics and chemistry together in the early 20th century.

This might seem obvious, but everything important in science got discovered out of order. Let's start over and piece it together again, knowing what we now know.

III: THERE IS ONLY ENTROPY: RESTRUCTURING THE NARRATIVE

Superimposing entropy back to the beginning of the universe creates a possibility for a more coherent scientific narrative.

"Time" is a relational concept, regarding movement, between two objects. A clock keeps steady movement in a world where movement is often chaotic. Temperature is a more direct measurement of movement precisely because work creates heat. Therefore, time began with movement. Scientists can identify only one known state where there is no movement, and that is in a pure crystalline substance at 0 K (see the chapter "The Third Law of Thermodynamics, String Theory, and Model-Dependent Realism").

We might ask ourselves, then, not "why is there something rather than nothing?" but "why is the universe not 0 K?" The answer to that may simply be that there are more ways to be not 0 K than there are ways to be 0 K.

This must be said because, while it makes sense that if the universe is expanding then it must have once been closer together, it doesn't necessarily make sense to posit a single infinitely dense particle because that equation indicates heat, and if there's heat, then there is movement, and therefore time.

We might therefore imagine, with some statistical certainty, a dense particle that was cold (0 K) and not moving. If the universe is expanding, then it is always at peak size. However, if it is always trending toward disorder, then it is also always at peak entropy. Reversing the concept of entropy to a singularity makes more sense than reversing movement to a singularity for reasons to be explained.

The emission of a beta particle from a 0 K singularity (similar to Hawking radiation) can be seen as the first movement. This would be more in keeping with a modern understanding and prevents us from conjuring up mathematical models based off predictions (postdictions?) that get less likely to reject the null hypothesis the farther back they go.

Because the concept of a "big bang" is based on reversing motion, it requires that the physicists posit the idea that we can measure the movements of that explosion by putting a hypothetical Earth in rotation around a hypothetical sun and then setting that outside of the real movements of particles so that modern scientists can have the reference point of a "year." The universe does not allow us this hypothetical reference point. This is why reversing entropy to a singularity makes more sense. No reference point is needed.

In the beginning, there was radiation, which is entropy, which is movement, which is time. From that, what we call "energy" should be defined as one object's temporary capture of another object's entropy. The sun doesn't shine; rather, it is radiating away, and Earth captures this temporarily to create temporary order. Entropy fuels biological evolution but also ensures only temporary order in an organism, hence the pressures on life to replicate for a future generation with greater capabilities to capture entropy.

The pre-Socratics asked, "What is the fundamental nature of matter?," not "Why does everything decay?" This, plus Aristotle's ruminations on motion, framed the scientific enterprise in a particular way so that, in 1666, Newton focused on why the apple fell, not why it rotted. It does no violence to the inverse square law to say that Newton "discovered" that the law of large numbers can determine that atoms move away from each other in a force proportional to their mass. The inverse square law can just as easily be used to describe how entropy's movement allows for objects to separate. The law of large numbers, not discovered until 1713, when the Swiss mathematician Jakob Bernoulli worked out the equations, joins Newtonian physics with the quantum in a coherent way.

Newton's calculations just happened to be worked out in a particular place and point in time where Earth's mass was large enough to temporarily hold back the entropy of the apple on a macro scale. On a micro scale, the apple still radiates away, which is why it rots and decays over time. Had Newton known about entropy, he might have described the elongation of the force of entropy as objects separate rather than the gravitational force of objects coming together.

If enough entropy is borrowed from another decaying source, then the apple's entropy can escape. The reason this is hard to see is because the ratio of Earth to the apple is not really that skewed. If Earth were made as heavy as a black hole and the apple as small as a beta particle, then we would now accept the fact of its radiation.

The second law of thermodynamics is not linear and absolute in the sense that not every interaction causes radiation that can be detected, but in the aggregate, the interactions create heat. This means that exothermic reactions produce entropy, while endothermic reactions temporarily absorb the entropy from another interaction. It does no violence to Einstein's equation if we make the E stand for "entropy" rather than "energy." It makes more sense to say that nuclear physics releases entropy because that force does not become energy until it is temporarily captured by another object.

This makes everything clear, because we can think of the elements in the periodic table as being arranged based on their temporary resistance to entropy. When elements interact, they release heat just as surely as Carnot found that machines do. Humans evolved on the surface of Earth, absorbing a certain amount of entropy from the sun and becoming resistant to most of the radiation (entropy) in the objects on the planet's surface. If we come into contact with materials that radiate faster that what our cells are used to, this can tear away DNA and cause defects and cancers.

What caused confusion at the dawn of quantum physics was the fact that equations in both physics and chemistry were based on the macro level, where the law of large numbers creates a ratio disparity between particles that makes entropy unimportant for temporary purposes. It doesn't matter, really, if there is some small level of heat loss that occurs when sodium and chloride are combined as long as table salt results. Stoichiometry ensures a balanced equation that is practical to use but not exactly accurate. The balanced equation on the right is a little less than the balanced equation on the left. The same is true of how large objects, composed of aggregated particles, act in relation to each other.

But quantum mechanics is the study of the radiated particles, and, individualized, the law of large numbers no longer applies to their behavior. This is why, as Heisenberg discovered, quantum equations are not commutative. If there is an interaction, then there is heat loss, and that cannot be reversed. (In a real way, this is measurable, as the paper you write your equation on is slightly hotter after the writing than it was before.) Matrix algebra allows for the quantum physicist to equate the position of a particle to a state of heat loss based on the interaction, but that cannot be reversed.

This is why work creates heat but heat does not create work. The second law is not commutative because time cannot go backward overall; entropy can be borrowed temporarily to reverse some processes, but, overall, everything decays.

Entropy keeps the arrow of time moving; today is less ordered than yesterday, and this is certain. If we extrapolate this concept backward through our scientific narrative to the origins of the universe, then we must postdict a universe that was once ordered only through its lack of movement, which means it was frozen. But even then, as Galileo once said of Earth, eppur si muove ("but it does move"). And if it moves, it creates heat, and understanding that creates a more coherent scientific narrative.

Appendix 2

*"Probability and Waves: A Synthesis of Three New Books"**

STATISTICS AND THEORETICAL PHYSICS, INTIMATELY CONNECTED, recently went through a forced update created by skepticism about the use of traditional explanatory models. The results of these updated theories are detailed in three new books: *Bernoulli's Fallacy: Statistical Illogic and the Crisis of Modern Science* (2021) by Aubrey Clayton, *Everything is Predictable: How Bayesian Statistics Explain Our World* (2024) by Tom Chivers, and *The Biggest Ideas in the Universe: Quanta and Fields* (2024) by Sean Carroll. Together they synthesize to create connections between Bayesian probabilities and wave function theories. This means that statisticians and physicists can model wave functions with probability functions more tightly than ever before, and in so doing avoid the logical problems that come from human-constructed analogies. Crucially, this is all coming just in time for AI to take the process over.

In *Bernoulli's Fallacy*, Aubrey Clayton writes about how the Swiss mathematician Jacob Bernoulli (1655–1705) created what became known as the frequentist method of statistical analysis. Clayton begins bluntly by stating, "The methods of modern statistics—the tools of data analysis routinely taught in high schools and universities, the nouns and verbs of the common language of statistical inference spoken in research labs and written in journals, the theoretical results of thousands of person-years' worth of effort—are founded on a logical error" (p. 1).

*This article by Chris Edwards was originally published online by *Skeptic* magazine on August 30, 2024.

Bernoulli drew a direct connection between large sample sizes and theoretical predictability. "What he was able to show," Clayton writes, "was that . . . observed frequencies would necessarily converge to true probabilities as the number of trials got larger" (p. 27). This concept stuck because it made intuitive sense. Yet, Bernoulli's frequentist approach suffered from a theoretical constraint. If a statistician predicts that "heads" will come up 25 times in 50 trials, the statistician can explain away a "40 heads to 10 tails" result by stating that the prediction and reality would align at a higher number of trials. "However," Clatyon writes, "this . . . presents a practical impossibility: we can't flip a coin an infinite number of times." Also, he adds, flip a coin that frequently and the coin will become misshapen in a way that a theoretical coin will not.

At some level, this is a common enough conceit. Anyone who has ever seen a "poll of polls" during an election season might have thought "what about a poll of poll of polls?" and realized that frequentist statistics might not converge on actual predictions in time for election night. Clayton then goes on to note that statistics hid the problem with p-values, an all-but-impossible-to-calculate concept about the predictive veracity of a sample. As is often the case in science, a concept overexplained is a concept flawed. As Clayton states, "In the centuries following Bernoulli, as empiricism began to crescendo, some made the leap to claiming this long-run frequency of occurrence was actually what the probability was by definition" (p. 29).

Untangle the p-value, and one finds only tautology. This is not unusual in theoretical mathematics, but statistical analysis matters more than theoretical math because statistical analysis of experiments often inform practical applications in public policy. Researchers can't just keep performing experiments over and over until theoretical predictions match experimental outcomes. Hence the problems that occur when experimental outcomes are not replicable. "These failures of replication," Clayton writes, "are compounded by the fact that effect sizes even among the findings that do replicate are generally found to be substantially smaller than what the original studies claimed" (p. 275). This is likely because only experiments with substantial "effect findings" are likely to get the attention and funding to be replicated, and the replication experiment gets a result that reverts to the mean.

So what do we do about the problem? Tom Chivers answers this question by channeling the theories of the English philosopher and statistician Thomas Bayes (1702–1761). Chivers writes that "for Bayes, probability is subjective. It's a statement about ignorance and our best guesses of the truth. It's not a property of the world around us, but of our understanding of the world" (p. 65). Later, Chivers contrasts Bayesian analysis with Bernoulli's approach by giving a nod to Clayton:

> The nature of frequentist statistics requires that you either reject the null or you don't. Either there's a real effect, or there isn't. And so, if you get a big enough sample size you'll definitely find something. A Bayesian, instead, can make an estimate of the size of the effect and give a probability distribution. (p. 149)

Because they are intuitive and practical, Bayesian probabilities don't need to hide any underlying flaws with p-values and null-hypotheses (although the latter has uses). The core Bayesian formula is simply understood as a variation of conditional probabilities: given what you know, what are the odds of a certain outcome? The more that you know, the more predictable your outcome can be. Conditional probabilities can be calculated by converging additive factors and then dividing those by singular factors that one has knowledge about, or the "given."

Late in the book, Chivers quotes a "superforecaster" named David Manheim, who treats statistical "base rates" (characteristics, by percentage, that exist in a population) as the core "given" in Bayesian analysis. While such an approach proves to be logically sound, it relies upon changing the mathematical probability analysis at the rate that new information (the "given") comes in. The problem, then, becomes practical because "most of us . . . don't keep base rates in our mind like that, so our beliefs are swayed by every new bit of information," writes Chivers. This leads to a new understanding of statistical analysis:

> . . . There's more to being a good forecaster than using base rates. For one thing, "using new data as a likelihood function" sounds nice and simple, but in most cases, you can't just do the math—yet people are

still using their judgement to decide how much to update from the base rate." (p. 258)

While this might sound subjective, Chivers advises forecasters to keep track of their predictions and measure them against each other, so that over time the predictions become less susceptible to erroneous base rates. In stating this, Chivers turns forecasting into a process informed first by intuitions about base rates but then shaped into specific predictions by adding more information to your Bayesian "given." To use an easy example, if a fifty-year-old woman walks into an oncologist's office; the oncologist would diagnose from a base rate about the percentages of cancer in the women-fifty-and-over category, but then shape the diagnosis with whatever data comes in from medical testing.

While Clayton and Chivers both approach statistical analysis as connected to sample size replication, diagnostics, and forecasting, their arguments must connect to the branch of physics most closely connected to probability: quantum mechanics. This brings us to Sean Carroll's book, dense with ideas and tightly packed with strings of logic: *The Biggest Ideas in the Universe: Quanta and Fields*. The book jacket promises the reader an explanation of theoretical physics that "goes beyond analogies." Reader beware: this means that Carroll requires that you stretch up to his explanations as he will not contract his theories to fit with what you might be more familiar with. There are no references to rubber bedsheets and bowling balls, and he tells you right away that nothing in the quantum world acts much like a game of billiards.

This reviewer found the approach to be refreshing more often than it was confusing. To take one example, Carroll never tries to describe anything by analogy to a "year," instead writing that "Nothing 'travels' through time, forward or backward; things exist at each moment of time and the laws of physics guarantee that there is some persistence of things from moment to moment. Certainly, any implication that we could use antiparticles to send signals backward in time, or somehow affect the past, is completely off base" (p. 107). Although Carroll does not explicitly say this, his approach sensibly does away with hypothetical time travel scenarios where we imagine a past state of the universe as compared to current measures of time.

Without analogies, and with incomplete information, physicists rely on probabilities. As Carroll notes, "Set up a quantum system with some wave function, let it evolve according to the Schrodinger equation, then calculate the probability of measurement outcomes using the Born rule. There's a bit of mathematical heavy lifting along the way, but the procedure itself is clear enough" (p. 41). Taking the measurement of a system causes it to break from a probabilistic potential reality into actual reality. Carroll writes that this is the measurement problem.

The introduction of a measurement force disrupts any connection of wave functions, Carroll later explains: "Because the apparatus is a macroscopic object that cannot help but interact with the bath of photons around it, the environment becomes entangled too. This process is known as decoherence—when a quantum system in a superposition becomes entangled with its environment" (p. 71). Carroll carefully states that a wave function is not a field. Later, he explains, "If the wave function is not a field, but we're doing quantum field theory, what is the relationship between the two? The answer is that we have a wave function of field configurations, in exactly the same sense that a quantum theory of particles features a wave function of particle positions" (p. 91).

Quantum physicists, and therefore Carroll, think of particles as statistical expressions of a field. There's not really a particle there, but a temporary knot of waves. Later, Carroll adds:

> Different fields have different spins, are characterized by different symmetries, and correspondingly come with their own notations . . . fields interact in myriad ways, and calculating the physical effects of those interactions can be a bit of a mess." (p. 98)

Connections with Bayesian probabilities can be found throughout Carroll's book, such as when he writes about the theoretical physicist and computer scientist Ken Wilson (1936–2013). In the 1960s Wilson was one of the first to see the direct application of computer technology to the problems of physics. This passage by Carroll provides the direct connection to the new understanding of Bayesian probabilities offered by Clayton and Chivers:

> Wilson was inspired by the newfangled device that was beginning to become useful to physicists: the digital computer. Imagine we try to do a numerical simulation of a quantum field theory. Inside a finite-sized computer memory, we cannot literally calculate what happens at every location in space, since there is an infinite number of such locations. Instead, we can follow the fields approximately by chunking space up into a lattice of points separated by some distance. What Wilson realized is that he could systematically study what happens as we take that separation distance to be smaller and smaller. And it's not just that we're doing calculus, taking the limit as the distance goes to zero. Rather, at any fixed lattice size we can construct an "effective" version of the theory that will be as accurate as we like. In particular, we can get sensible physical answers without ever worrying about—or even knowing—what happens when the distance is exactly zero. (p. 129)

Put another way, quantum particles move a lot faster than we do. By the time a human observer can cognitively process the position of a particle, the particle has moved on. Human observers can only see the universe as it was, never as it is. A computer can mathematically model a wave function by assessing new information as a Bayesian "given" and can recalculate probabilities from that about a particle's position faster than humans can. The distance between an event and our understanding of an event can never be zero, but computers can process information and break down probabilities with information at a rate that humans cannot comprehend.

SYNTHESIS AND ANALYSIS

Since Clayton and Chivers found something new by studying a couple of old mathematicians, it is worth dusting off one of James Clerk Maxwell's most underappreciated ideas regarding probabilities and temperature.

In *Convergence: The Idea at the Heart of Science* (2017), Peter Watson writes:

> Maxwell saw that what was needed was a way of representing many motions in a single equation, a statistical law. He devised one that said nothing about individual molecules but accounted for the proportion that had the velocities within any given range. This was the first-ever statistical law in physics, and the distribution of velocities turned out to be bell-shaped, the familiar normal distribution of populations about a mean. But its shape varied with the temperature—the hotter the gas, the flatter the curve and the wider the bell. (p. 39)

Maxwell's theory connects quantum mechanics with probability theory. From a probability perspective, an object at absolute zero creates a flat probability line because all the particles adhere directly to the mean. Adding heat to a system causes particles and waves to fly off into different probability states. As Carroll notes, "Heavier particles decay into lighter ones. In special relativity, energy is conserved but 'mass' is not separately conserved; it is just one form of energy. A heavy particle can decay into a collection of particles with a lower mass, with the remaining energy being the kinetic energy of the offspring particles" (p. 257). He just described entropy, or the heat loss that comes from interactions. Any moving system, and there is no other kind, releases heat that can be directly translated into Maxwell's probability field.

Vibrating strings can be understood, for example, in a direct ratio to the heat lost by the vibration. Cooling an object to absolute zero produces the same "slowing time" effect as time dilation, for example, which means that "time" as a concept can be understood as relative motion or as relative temperature, but the temperature model more easily fits with our understanding of probability. Carroll states that "no matter how big our system is, there is only one wave function. Ultimately, there is just one quantum state for everything, the wave function of the universe" (p. 63). That function can be broken up into smaller systems and understood through probability and this means that we can't get the absolute truth of wave function, but we can get pretty close.

To take Carroll's "no-analogies" conceit to its logical conclusion, computer-generated AI will not need the esoteric names that we assign to the elements. Each element can just as well be understood as being in a probabilistic state of entropy; with stable elements decaying slowly and unstable elements radiating quickly. Because computers can turn information into Bayesian probabilities so quickly, an AI can adjust its base rate accordingly and close the gap between "given" information and probabilities instantly. AI will go beyond human theoretical capacities quickly, but thanks to the theoretical updates, driven by the skepticism of old methods, given to us by Clayton, Chivers, and Carroll, we have a chance to understand how AI does it.

REFERENCES

Aristotle. (1999). *Metaphysics* (W. D. Ross, Trans.). Oxford University Press.

Bohm, D. (1951). *Quantum Theory*. Dover.

Bohr, N. (1961). *Atomic Physics and Human Knowledge*. Dover.

Born, M. (1921). *Problems of Atomic Dynamics*. Dover.

Burke, J. (1985). *The Day the Universe Changed: How Galileo's Telescope Changed the Truth and Other Events in History That Dramatically Altered Our Understanding of the World*. Little, Brown.

Cardano, G. (2019). *The Rules of Algebra* (T. R. Witmer, Trans.). Dover.

Carnot, S. (2005). *Reflections on the Motive Power of Fire, and on Machines Fitted to Develop That Power*. Dover.

Carroll, Sean. (2024). *The Biggest Ideas in the Universe: Quanta and Fields*. Penguin.

Chivers, Tom. (2024). *Everything Is Predictable: How Bayesian Statistics Explain Our World*. One Singal Publishers.

Clark, C. (2006). *Iron Kingdom: The Rise and Downfall of Prussia, 1600–1947*. The Belknap Press of Harvard University Press.

Clausius, R. (1879). *On the Motive Power of Heat, and on the Laws Which Can Be Deduced from It for the Theory of Heat*. Macmillan.

Clayton, Aubrey. (2021). *Bernoulli's Fallacy: Statistical Illogic and the Crisis of Modern Science*. Columbia University Press.

Cohen, S. M., Curd, P., and Reeve, C. D. C. (Eds.). (2016). *Readings in Ancient Greek Philosophy, from Thales to Aristotle* (5th ed.). Hackett.

Crosby, A. W. (1997). *The Measure of Reality: Quantification and Western Society, 1250–1600*. Cambridge University Press.

Darwin, C. (1859). *On the Origin of Species by Means of Natural Selection, or the Preservation of Favoured Races in the Struggle for Life*. Penguin Classics.

Dawkins, R. (1999). *The Extended Phenotype: The Long Reach of the Gene*. Oxford University Press.

Dennett, D. C. (1995). *Darwin's Dangerous Idea: Evolution and the Meanings of Life*. Simon & Schuster.

Dictionary of Scientific Theories. (2001). Greenwood Press.

Ditchburn, R. W. (1961). *Light*. Dover.

DK. (2012). *Physics*. DK.

Edmonds, D., and Eidinow, J. (2001). *Wittgenstein's Poker: The Story of a Ten-Minute Argument between Two Great Philosophers*. HarperCollins.

Edoardo, A. (2020). *All about History—The Templars: The Rise and Fall of the Secretive Military Order*. Future Publishing Ltd.

Edwards, C. (2011). "Baptism by Fire: Christian Violence, Sources." *Skeptic*. Retrieved from https://www.skeptic.com/reading_room/how-christian-violence-abetted-advent-of-nationalism/.

———. (2013). "Baptism by Fire: Christian Violence." *Skeptic*. Retrieved from https://www.skeptic.com/reading_room/how-christian-violence-abetted-advent-of-nationalism.

———. (2014). "Illuminated Genius." *Skeptic*. Retrieved from https://www.skeptic.com/reading_room/illuminated-genius-einsteins-masterwork-general-theory-relativity.

———. (2017). "Upending Civilization." *Skeptic*. Retrieved from https://www.skeptic.com/reading_room/upending-civilization-review-dawn-of-everything.

———. (2021). *Femocracy: How Educators Can Teach Democratic Ideals and Feminism*. Rowman & Littlefield.

Forbes, N., and Mahon, B. (2006). *Faraday, Maxwell, and the Electromagnetic Field: How Two Men Revolutionized Physics*. Prometheus Books.

Fowles, G. R. (1975). *Introduction to Modern Optics* (2nd ed.). Dover.

Gamow, G. (1985). *Thirty Years That Shook Physics: The Story of Quantum Theory*. Dover.

———. (2005). *The Birth and Death of the Sun*. Dover.

Gödel, K. (1931). *On Formally Undecidable Propositions of Principia Mathematica and Related Systems*. Dover.

Graeber, D. (2018). *Bullshit Jobs: A Theory*. Simon & Schuster.

Graeber, D., and Wengrow, D. (2021). *The Dawn of Everything: A New History of Humanity*. New York: Farrar, Straus and Giroux.

Greenblatt, S. (2011). *The Swerve: How the World Became Modern*. Norton.

Gribbin, J. (1996). *Schrödinger's Kittens and the Search for Reality: Solving the Quantum Mysteries*. Little, Brown.

———. (2003). *The Scientists: A History of Science Told through the Lives of Its Greatest Inventors*. Random House.

———. (2016). *Einstein's Masterwork: 1915 and the General Theory of Relativity*. Pegasus Books.

Hawking, S. W. (1988). *A Brief History of Time*. Bantam.

Hawking, S. W., and Mlodinow, L. (2010). *The Grand Design*. Bantam.

Hecht, J. (2004). *Doubt: A History*. HarperOne.

Hidden Games: Human Choices and Systems Behavior. (2001). Chelsea Green Publishing.

Hollingdale, S. (2011). *Makers of Mathematics*. Dover.

Ihde, A. J. (2012). *The Development of Modern Chemistry*. Dover.

Kirsch, J. (2004). *God against the Gods: The History of the War between Monotheism and Polytheism*. Viking.

Kline, M. (1967). *Mathematics for the Nonmathematician*. Dover.

———. (1998). *Calculus: An Intuitive and Physical Approach*. Dover.

Kuhn, T. S. (1962). *The Structure of Scientific Revolutions*. University of Chicago Press.

Kumar, M. (2011). *Quantum: Einstein, Bohr, and the Great Debate about the Nature of Reality*. Norton.

Labatut, B. (2023). *The Maniac* (M. Hulse, Trans.). Penguin.

Lakoff, G., and Johnson, M. (1980). *Metaphors We Live By*. University of Chicago Press.

Lavenda, Bernard H. (1978). *Thermodynamics of Irreversible Processes*. Dover.

Lavoisier, A. (1789). *Elements of Chemistry*. Dover.

Leslie, J., and Kuhn, R. L. (Eds.). (2013). *The Mystery of Existence: Why Is There Anything at All?* Wiley.

Lightman, A. (2005). *The Discoveries*. Vintage.

Lucretius. (50 BC). *On the Nature of Things* (M. F. Smith, Trans.). Penguin Classics.

McCarthy, C. (1992). *Blood Meridian: Or the Evening Redness in the West*. Vintage.

———. (2022). *The Passenger*. Knopf.

Medieval Philosophers: An Anthology. (2003). Hackett.

Muller, R. A. (2016). *Now: The Physics of Time*. Random House.

Myklestad, N. O. (2013). *Fundamentals of Vibration Analysis*. Dover.

Nash, L. K. (1968). *Elements of Chemical Thermodynamics* (2nd ed.). Dover.

Newman, J. R. (Ed.). (1956). *The World of Mathematics* (Vols. 1-4). Simon & Schuster.

Norwich, J. J. (2011). *The Popes: A History*. Random House.

Ore, O. (1948). *Number Theory and Its History*. Dover Publications.

Planck, M. (1914). *The Theory of Heat Radiation*. Dover Publications.

Plutarch. (1914). *Plutarch Lives, II: Themistocles and Camillus. Aristides and Cato Major. Cimon and Lucullus* (B. Perrin, Ed.). Harvard University Press.

Price, D. A. (2003). *Love and Hate in Jamestown: John Smith, Pocahontas, and the Heart of a New Nation*. Vintage.

Reston, J. (2005). *Dogs of God: Columbus, the Inquisition, and the Defeat of the Moors*. Anchor Books.

Rovelli, Carlo. (2022). *Helgoland: Making Sense of the Quantum Revolution*. Penguin Random House.

Russell, B. (1946). *A History of Western Philosophy*. Routledge.

Sacks, O. (2017). *Exact Thinking in Demented Times: The Vienna Circle and the Epic Quest for the Foundations of Science*. Basic Books.

Sarma, Sanjay E., Steven Schofield, James Stori, Jane MacFarlane, and Paul K. Wright. (1996). "Rapid Product Realization from Detail Design." *Computer-Aided Design* 28, no. 5 (May 1996): 383–92.

Sawyer, W. W. (2007). *Mathematician's Delight*. Dover.

Schlosser, E. (2013). *Command and Control: Nuclear Weapons, the Damascus Accident, and the Illusion of Safety*. Penguin Books.

Segre, E. (1980). *From X-Rays to Quarks: Modern Physicists and Their Discoveries*. Dover.

Seife, C. (2016). *Hawking: The Selling of a Scientific Celebrity*. Basic Books.

Shlain, L. (1998). *The Alphabet versus the Goddess: The Conflict between Word and Image*. Penguin.

Spielvogel, J. J. (2019). *Western Civilization*. Cengage Learning.

Tolman, R. C. (2011). *Relativity, Thermodynamics and Cosmology*. Dover.

Watson, P. (2017). *Convergence: The Idea at the Heart of Science*. Simon & Schuster.

———. (2001). *The Modern Mind: An Intellectual History of the 20th Century*. Harper Perennial.

———. (2010). *The German Genius*. Harper Perennial.

Weatherford, J. M. (2005). *Genghis Khan and the Making of the Modern World*. Crown.

Wikipedia contributors. (n.d.). "Law of Large Numbers." Retrieved from https://en .wikipedia.org/wiki/Law_of_large_numbers.

———. (n.d.). "Rudolf Clausius." Retrieved from https://en.wikipedia.org/wiki/Rudolf _Clausius.

Wilson, E. O. (1999). *Consilience: The Unity of Knowledge*. Vintage.

Wilson, J. (1999). *God's Funeral: The Decline of Faith in Western Civilization*. Norton.

Wittgenstein, L. (1953). *Philosophical Investigations*. Blackwell.

Wittgenstein, L. (2009). *Major Works: Selected Philosophical Writings*. Harper Perennial.

Wollstonecraft, M. (1792). *A Vindication of the Rights of Women* (D. Todd, Ed.). Penguin Classics.

Wootton, D. (2016). *The Invention of Science: A New History of the Scientific Revolution*. Harper Perennial.

About the Author

Chris Edwards, EdD, teaches history and English at a public school in the Midwest. He is a frequent contributor to *Skeptic* magazine and the author of numerous books on philosophy, science, world history, and educational theory. His original "connect-the-dots" teaching method was published by the National Council for Social Studies. Chris is a former research affiliate with the Massachusetts Institute of Technology.